Basic Concepts in Relativistic Astrophysics

Basic Concepts in Relativistic Astrophysics

Li Zhi Fang

Astrophysics Research Division
University of Science & Technology of China

Remo Ruffini

Instituto di Fisica G Marconi
Università degli Studi — Roma

World Scientific

World Scientific Publishing Co. Pte. Ltd.
P O Box 128
Farrer Road
Singapore 9128

ISBN 9971-950-66-9

Printed in Singapore by Singapore National Printers (Pte.) Ltd.

Preface

The idea of writing this book started when one of us (R Ruffini) visited China in May 1978. We discovered that we have carried out quite a few similar works independently. (For example, we arrived using different methods, at the values of $3.2M_\odot$ and $3.18M_\odot$ for the critical mass of a neutron star.) We then decided to write this book as a token of the exchange between the East and the West. A rough content was drafted during trips to Suzhou, Hangzhou and Guangzhou.

The other author (L Z Fang) visited Rome during March and April of 1979 and the first draft of this book was written during our visits to the Tower of Pisa, Florence and Venice.

It is hoped that this book will lead to greater cooperation between the scientists in the East and the West.

L Z Fang
R Ruffini

CONTENTS

Preface   v

Chapter 1: BASIC CONCEPTS OF GENERAL RELATIVITY

    1.1.   The Aims and Characteristics of Astrophysics   1
    1.2.   Newtonian Mechanics and Absolute Space   3
    1.3.   The Difficulties of Newtonian Gravitation   6
    1.4.   Special Relativity and Gravity   10
    1.5.   Mach's Principle   12
    1.6.   Principle of Equivalence   14
    1.7.   Local Inertial Frames   17
    1.8.   Metrics of Spacetime   19
    1.9.   Geodesics   23
    1.10. Curved Spaces   26
    1.11. Relation between Curvature and Matter   33
    References   38

Chapter 2: EFFECTS IN A WEAK GRAVITATIONAL FIELD

    2.1.   Gravitational Redshift   40
    2.2.   Schwarzschild Metric   44
    2.3.   Clocks Moving Around the Earth   46
    2.4.   Precession of the Perihelion   49
    2.5.   Deflection of Light   56
    2.6.   Experiments on Radar Echoes   61
    2.7.   Precession of the Axis of Rotation   65
    References   70

Chapter 3: COMPACT STARS

    3.1.   Historical Records   71
    3.2.   Crab Nebula   77

3.3.   Theory of Cold Stars   83
3.4.   Neutron Stars   90
3.5.   The Discovery of Pulsars   96
3.6.   Basic Properties of Pulsars   103
References   110

Chapter 4: BLACK HOLES
4.1.   Critical Mass   111
4.2.   The Phenomena in a Collapse   114
4.3.   The Types of Black Holes   121
4.4.   Horizons and Emissions from Black Holes   123
4.5.   Identification of Black Holes   129
4.6.   X-ray Pulsars in Close Binary Star Systems   135
4.7.   Close X-ray Binary Star Systems with Fluctuating
       Luminosities   140
References   143

Chapter 5: GRAVITATIONAL WAVES
5.1.   Gravitational Fields of 1/r Type   144
5.2.   Deviation Equation   147
5.3.   Essential Properties of Gravitational Waves   148
5.4.   The Radiation of Gravitational Waves   151
5.5.   Gravitational Radiation of Binary Stars   153
5.6.   Observational Verification of Gravitational
       Radiation Damping   156
References   164

Chapter 6: RELATIVISTIC COSMOLOGY
6.1.   Difficulties between the Infinite Universe and
       Newtonian Theory   165
6.2.   Spacetime of a Homogeneous and Isotropic Universe   167
6.3.   Relations between Apparent Magnitudes and Red-Shifts   170
6.4.   The Darkness of the Night Sky   178
6.5.   Number Counts   180
6.6.   Dynamical Equation for R(t)   183
6.7.   Age of the Universe   187

6.8.   Background Black Body Radiation   193

6.9.   Helium Abundance   201

6.10.   Formation of Galaxies   205

6.11.   The Very Early Universe   211

References   214

Index   215

Chapter 1

BASIC CONCEPTS OF GENERAL RELATIVITY

1-1   The Aims and Characteristics of Astrophysics

To deal generally with the very large scale of spacetime is one
of the aims of astrophysics. The reception of light and its emission
from distant galaxies are met most generally and the relationships
between these events, separated by great distances, are often discussed
in astrophysics.

In order to describe events, we need a frame of reference, within
which events are denoted with respect to a coordinate system. Each
event is specified by four numbers, usually taken as three position
coordinates in $\vec{r}$ and one time coordinate t.

The most convenient frame of reference for describing the reception of light is a frame at rest with respect to us or to our own
galaxy, or to the centre of mass of the local group of galaxies. On
the other hand, the simplest frame for describing the emission of light
from a distant galaxy G is a frame at rest with respect to it or to
the centre of mass of the group of galaxies within which the galaxy G
is located. In this frame, the emission event has coordinates $\vec{r}'$, t'.
Evidently, a question which must be answered is how we can transform
the coordinates of an event in one frame to get the coordinates in

another. It is only by solving this problem that we can find the coordinates $\vec{r}$, t of this emission event in our own frame.

According to Newtonian mechanics, a transformation between the
so-called inertial frames is a Galilean transformation given by the
well known formulas as follows:

$$x' = x - v t \quad,$$
$$y' = y \quad,$$
$$z' = z \quad,$$
$$t' = t \quad,$$

where (t,x,y,z) are the coordinates of an event in an inertial frame
K and (t',x',y',z') the coordinates of the same event in another
inertial frame K', moving relative to K with uniform velocity v
along the positive x-axis of K.

Unfortunately, in astrophysics, Galilean transformations cannot
generally be employed because of two reasons. Firstly, in astrophysics,
high speed situations are frequently encountered, such as the emission
of light from a galaxy G and the reception of this light by us, as
discussed above. If G is distant enough from us, then, generally
speaking, it has a very large speed relative to us, and the transformations which can only be applied with validity to low speed cases cannot
be employed. Secondly, our galaxy and G may be accelerating mutually
under gravity, and it is therefore necessary to find a new transformation which is valid in the presence of gravity.

The second point mentioned above is characteristic of astrophysics: only in the problems of astrophysics is gravity dominant. In
contrast, in the physics of a laboratory on the earth, gravity is less
important.

Gravity is the weakest of the fundamental interactions. The
ratio of the electrostatic to gravitational forces between an electron
and a proton is about $e^2/Gm_e m_p \approx 10^{39}$, which is why we neglect gravitation in atomic physics. This is also true within nuclei.

Gravity dominates all other forces when considered on the astro-

2

nomical scale since gravity is a long-range force, whereas nuclear forces are negligible on the large scale, and interatomic forces such as the Van der Waal's forces falling off roughly as $1/r^7$, are negligible for two atoms separated by more than about 10^{-9} m. As a result, these strong short-range forces are insignificant in systems widely separated in a vacuum, compared with gravitational and electrostatic forces. Although the electrostatic force is long-range, it can be "screened". Electric charges can be positive or negative, but a celestial body as a whole is neutral, so that all long-range electrostatic forces are ineffective. Gravity is very different from the electrostatic force in that it cannot be screened, since all material bodies attract each other.

It is inevitable that problems on a large scale in spacetime should encounter gravity and a correct theory of gravitation is necessary for relating the events on the astronomical scale. Newton's theory of gravitation is quite good in the case of weak gravitational fields and at low speeds, but the situations discussed in relativistic astrophysics often involve strong gravitational fields, or high speeds, or both at the same time. These problems are peculiar to and characteristic of astrophysics.

1-2   **Newtonian Mechanics and Absolute Space**

Newton's second law, which is the nucleus of the whole of Newtonian mechanics, is represented most generally as

$$\vec{F} = m\,\vec{a} \;, \tag{1.1}$$

where m is the mass of the body, $\vec{a}$ is its acceleration relative to a certain frame of reference, and $\vec{F}$ is the resultant force acting on the body due to all other bodies.

The types of frames in which Newtonian mechanics can be employed are called inertial frames, moving relative to each other with uniform velocities. It follows that the inertial frames occupy a special position in Newtonian mechanics. We wish to ask more fundamental questions like why Nature selects such frames preferentially and how does one determine whether a given frame is inertial. It is generally agreed that inertial frames are unaccelerated and non-rotating. If they are not, we are then led to ask what the acceleration and rotation are relative to.

Newton himself had given his answers to these questions. The key to his answer lies in the introduction of the concept of absolute space, that is, inertial frames are unaccelerated and non-rotating relative to absolute space, which "in its own nature, without relation to anything external, remains always similar and immovable". According to him, acceleration and rotation relative to absolute space are detectable by simple experiments.

For example, if we accelerate forwards, we see the room accelerate backwards. As there is actually no backward force acting on the room, according to Newton, the apparent acceleration of the room is not a consequence of Newton's second law, but of our acceleration relative to absolute space.

As another example, consider a rotating pail of water. Is its rotation absolute (i.e. relative to absolute space), or is it a relative rotation (equivalent to what we perceive when we ourselves are actually rotating)? Newton's criterion is that we should look and see whether the surface of the water appears to be concave; if it does, then the pail of water is rotating relative to absolute space, and if not then the converse is true.

However, although Newton could give the criterion for absolute acceleration and absolute rotation, he could not do so for absolute velocity. In fact, even before Newton's time it was already realized that it is impossible to find a criterion for "absolute rest". Two thousand years ago, a paragraph of the ancient Chinese text "Shang-Shu-Wei. Kao-Ling-Yao" was written as follows: "The earth is constantly in motion but we do not know it. This is like people sitting in a boat with all the windows closed; the boat moves but those inside feel nothing". Galileo's exposition shows the same idea as we shall see in the following excerpt from his "Dialogues Concerning Two New Sciences".

*Shut yourself up with some friend in the main cabin below deck
on some large ship, and have with you there some flies, butterflies and
other small flying animals. Have a large bowl of water with some fish
in it; hang up a bottle that empties drop by drop into a wide vessel
beneath it. With the ship standing still, observe carefully how the
little animals fly with equal speed to all sides of the cabin. The
fish swim indifferently in all directions; the drops fall into the
vessel beneath; and, in throwing something to your friend you need not
throw it more strongly in one direction than another, the distances
being equal; jumping with your feet together, you pass equal spaces in
every direction. When you have observed all these things carefully
(though there is no doubt that when the ship is standing still every-
thing must happen in this way), have the ship proceed with any speed
you like, so long as the motion is uniform and not fluctuating this way
and that. You will discover not the least change in all the effects
named, nor could you tell from any of them whether the ship was moving
or standing still. In jumping, you will pass on the floor the same
spaces as before, nor will you make larger jumps towards the stern than
towards the prow even though the ship is moving quite rapidly, despite
the fact that during the time you are in the air the floor will be going
in a direction opposite to your jump. In throwing something to your
companion, you will need no more force to get it to him whether he is
in the direction of the bow or the stern, with yourself situated opposite.
The droplets will fall as before into the vessel beneath without dropping
towards the stern, although while the drops are in the air the ship runs
many spans. The fish in their water will swim towards the front of their
bowl with no more effort than towards the back, and will go with equal
ease to bait placed anywhere around the edge of the bowl. Finally, the
butterflies and flies will continue their flights indifferently towards
every side, nor will it ever happen that they are concentrated towards
the stern as if tired out from keeping up with the course of the ship,
from which they will have been separated during long intervals by keeping
themselves in the air.*

Summing up the above, it follows that absolute velocity cannot be measured. This is the first objection to the concept of absolute space.

Secondly, Newton's absolute space is a physical reality, i.e. a structure with physical influence, which has an effect upon dynamical properties of matter. For example, it is a very strong effect that the law of inertia holds only relative to absolute space. However, matter does not act on it, because it is without relation to anything external. This one-way relationship, i.e. the existence of an action without its reaction, is not satisfactory in physics.

The third objection to absolute space comes from observational experiments. As we know, the Earth is a good approximation to an inertial frame and our observations of phenomena in the laboratory are always relative to it. It appears in the dynamics of the planets that the frame in which the Sun is at rest relative to the average motion of the nearby stars is closer to the inertial frame than the Earth, and we do not use the latter but the former for describing the motion of the planet. By further studying the dynamics of the galaxies, as we have already found, a better inertial frame is one in which the centre of the galaxy appears at rest relative to the average motion of other galaxies. In this way, we can get progressively closer and closer to an ideal inertial frame; that is, the larger the scale of the astronomical system in which the frame appears at rest relative to the average motion, the closer to an ideal inertial frame. Thus it can be seen that the ideal inertial frame is at rest relative to the average motion of matter in the universe. Since the universe is expanding, an equivalent specification of this "ultimate frame" is that from it the expansion should appear isotropic. By this empirical inference, it can be shown that Newton's absolute space, i.e. the most ideal and preferred inertial frame, is not "without relation to anything external" but tied to the distribution and motion of matter on a large scale.

1-3   The Difficulties of Newtonian Gravitation

In Newtonian gravitation, the mutual attraction between two particles of masses m_1 and m_2 separated by a distance r is (Fig. 1.1)

6

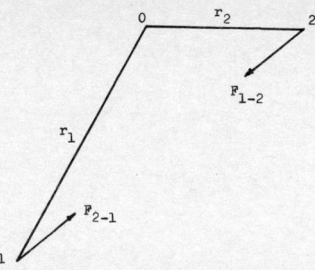

Fig. 1.1 Newtonian Theory of Gravitation

$$F = G \frac{m_1 m_2}{r^2} \quad , \tag{1.2}$$

where G is the gravitational constant, its value being 6.67×10^{-11} $Nm^2 kg^{-2}$.

Although Newton's theory gives a correct quantitative description of the gravitational force, the most elementary feature of gravitation is still not well defined. Which feature of gravitation is then the most important, if we were to consider the most fundamental?

By comparing Newton's second law (1.1) with his law of gravitation, we can describe the motion of a freely falling object by using the following equation:

$$m_i \vec{a} = m_g \frac{GM_\oplus}{r^3} \vec{r} \quad , $$

where m_i and m_g represent respectively the object's inertial mass (inversely proportional to acceleration) and the gravitational mass (directly proportional to gravitational force), $M_\oplus$ is the gravitational mass of the Earth, and $\vec{r}$ is the position vector of the object taken from the centre of the Earth. The above equation can be rewritten as

$$a = \frac{m_g}{m_i} \left(\frac{GM_\oplus}{r^2} \right) \quad . \tag{1.3}$$

7

Galileo's Pisa experiment showed that irrespective of the object chosen, the acceleration of an object produced by the gravitational force is the same, which from eq. (1.3) implies that the value of m_g/m_i should be the same for all objects. In other words, we have

$$m_g/m_i = \text{universal constant.} \tag{1.4}$$

In 1889, Eötvös showed that m_g/m_i was unity to within 10^{-9} for wood and platinum, while Dicke and others in 1964 showed that aluminium and gold fall towards to the sun with the same acceleration, to one part in 10^{-11}. Nordtvedt considered the possibility of violating the universal value of m_g/m_i by introducing a coefficient η to represent such a violation:

$$m_g = m_i (1 + \eta \Delta) \quad .$$

Laser measurements of the Earth-Moon separation are sufficiently sensitive to detect the difference between Δ(Earth) and Δ(Moon). The observable effect for $\eta > 0$ would be a displacement towards the Sun of the orbit of the Moon around the Earth. The corresponding change in the Earth-Moon distance would have a period of a month and an amplitude of about 8η metres. From an analysis of some 2000 of the lunar laser "normal points", obtained after about 5 years of observation, the estimate for η was consistent with zero.

In physics, the discovery of a universal constant often leads to the development of an entirely new theory. From the universal constancy of the velocity of light c, the special theory of relativity was derived; and from Planck constant h, the quantum theory was deduced. Therefore, the universal constant m_g/m_i should be the key to the gravitational problem. The theoretical difficulty with Newtonian gravitation is to explain just why relation (1.4) exists implicitly in Newton's theory as a separate law of nature besides (1.1) and (1.2).

Even so, we must say that Newtonian gravitation and mechanics was the first truly successful dynamics, and its most well-known application was in celestial mechanics. The verification of the prediction of the existence of Neptune marked the peak of the success of celestial mechanics but the first real difficulty was also met here. It was first pointed out in 1850, based on astronomical observations, that there was a discrepancy between certain observations of the orbit of Mercury and the predictions made by Newtonian mechanics. According to Newton's theory of gravitation, the Sun's gravitational force acting on Mercury causes its orbit to be a closed ellipse (Fig. 1.2). In fact it is not a precise ellipse: with every revolution, its major axis rotates slightly (Fig. 1.3). The rotation of the major axis is called the precession.

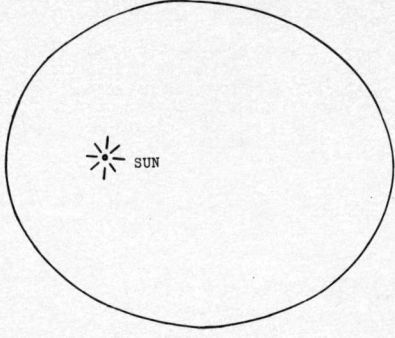

Fig. 1.2 Elliptical orbit of planets
under the gravity of the Sun

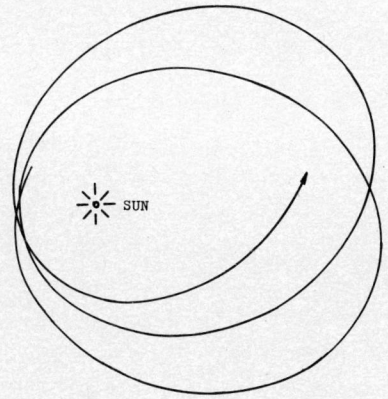

Fig. 1.3 Precession of the perihelion of Mercury

9

The observed rate of Mercury's precession is $1^{o}33'20''$ per century. This value ought to be due to the gravitational perturbations of all other planets and the effect of rotation of our Earth-based coordinate system. However, the value calculated from Newtonian mechanics is $1^{o}32'37''$ per century. The discrepancy between them of

$$1^{o}33'20'' - 1^{o}32'37'' = 43''$$

is extremely small, but it has been observed with a negligible amount of observational error.

The 43" problem has given rise to a great deal of debate. Le Verrier, who had successfully predicted the existence of Neptune, turned to this old sesame again and predicted the presence of a small planet near the sun which would account for the unusual precession. This time Le Verrier failed. No new planet was found at the place and time as predicted. In a few words, despite the fact that 43" per century is rather small, it actually represents a tremendously outstanding problem for Newtonian mechanics.

1-4   Special Relativity and Gravity
Having established electrodynamics, Maxwell predicted that electro-magnetic waves which travel with the speed of light c should exist, but what is this speed c with respect to? With respect to the ether was the initial answer. The ether was the medium "in which" electro-magnetic waves were supposed to travel, rather like air is the medium in which sound waves travel.

After proposing the concept of the ether, it was natural to iden-tify the ether with Newton's absolute space; this would unify electro-magnetism with mechanics. Experiments were carried out with the aim of measuring the Earth's velocity through the ether, i.e., measuring the velocity of absolute motion, by electromagnetic methods, and hence determine the absolute space. If one imagines the Earth as a ship floating in the ether, the speeds of light which propagate in different directions on the Earth should differ. On the contrary, the experiments showed that such differences do not exist. Perhaps the Earth drags

10

nearby ether along with it, so that terrestrial experiments could not reveal any directional dependence of the speed of light. However, this also is untenable, since it contradicts experiments on aberration.

These experiments, in fact, further support the inference made in section 1-2: absolute velocity cannot be found by electromagnetic or optical methods; or in other words, all inertial frames are equivalent. This idea was introduced as the relativity principle by Einstein and was stated in his article on special relativity in 1905 as:

> *The laws by which the states of physical systems undergo change are not affected, whether these changes of state be referred to the one or the other of two systems of coordinates in uniform translatory motion.*

This is one of two main principles of special relativity. The other is the principle of the constancy of the velocity of light, i.e. the velocity is always equal to c and is independent of the motion of the source of light.

Using these principles, the coordinate transformation called the Lorentz transformation can be derived. This concerns two frames K and K', with the origin of K' moving relative to K with speed v along the positive x-axis of K. At time $t = t' = 0$, K and K' are coincident. For an event whose K coordinates are (x,y,z,t), the K' coordinates are (x',y',z',t') given by

$$x' = \frac{(x - vt)}{\sqrt{1 - (v/c)^2}} \quad ,$$

$$y' = y \quad ,$$

$$z' = z \quad ,$$

$$t' = \frac{(t - vx/c^2)}{\sqrt{1 - (v/c)^2}} \quad . \tag{1.5}$$

The Lorentz transformation tends to the Galilean transformation mentioned in section 1-1 when the velocity of light c can be considered infinite, i.e. $c \to \infty$.

11

One of the greatest differences between the Galilean and the Lorentz transformation is that time is absolute in the Newtonian system; that is, two events which are simultaneous as observed in an inertial system are also simultaneous in another. In contrast, such absolute simultaneity is abandoned by the spacetime viewpoint of Einstein.

Special relativity does unify electromagnetism with mechanics on new ground and shows that if absolute space exists it cannot be identified with the classical ether. However, special relativity does not completely solve the problems of Newtonian theory discussed in section 1-3. First of all, the inertial frames are still in a "superior" position in special relativity, and we do not know why the frame at rest relative to the average frame in the universe is inertial.

The problems of gravity are also unsolved within the context of special relativity. The anomaly in the centennial precession of Mercury remains. One of the postulates of special relativity is that the velocity of light in both vacuum and in the presence of electromagnetic fields is the constant c for all inertial observers (if the small effect of quantum electrodynamics can be neglected). This may lead to the conclusion that the velocity of light is still constant c in a gravitational field, by using some theory which tries to graft gravity onto special relativity. This conclusion is wrong and contradicts observations.

1-5   Mach's Principle

Newton's absolute space was abandoned by special relativity due to the limitations of inertial frames, that is, because we cannot identify the inertial frame which is at rest relative to absolute space. However, it seems to find clear differences between inertial frames and non-inertial frames. For example, Newton's pail of water can identify absolute acceleration.

Right from the start, Leibnitz and Berkeley had disagreed with Newton about absolute space. Their argument arose not only from the non-possibility of measuring absolute velocity, but also from the

12

incorrectness of the concepts of absolute acceleration and absolute rotation. Mach developed this point of view in a more consistent manner.

Newton asserted that we can imagine the existence of a space and determine a body's acceleration relative to the space even if matter does not exist in the universe. However, consider a universe containing only two bodies and ask the question of whether they are rotating about their centre of mass. Newton would argue that this can be decided by connecting the bodies with a spring; if the system had an absolute rotation the spring would be extended, and the absolute rotation would be identified by this observable extension of the spring. Leibnitz and Berkeley would disagree with this and assert that there could never be any observed extension of the spring in such a universe of only two bodies. They would claim that Newton's argument about the imaginary space is based on his observations of the real universe. Nevertheless, there is a tremendous difference between the imaginary universe and the real universe, where a huge conglomerate of stars and galaxies provide a background. In the latter case, while the extension of the spring would be observed, the system of two bodies would be rotating relative to the huge collection of galaxies and not relative to the absolute space, the concept of which is useless.

In 1872, Mach asserted a case that the above is not absolute acceleration but acceleration relative to the distant galaxies. There is a simple experiment which makes this argument very persuasive. We stand outside on a starry night and look up. The stars are at rest and our hands under natural circumstances would be extended downwards. When we suddenly pirouette as fast as we can, two remarkable things happen: the stars rotate, and our arms rise up and point outwards. In Newton's view, the two phenomena are unconnected and the rising-up of our arms comes from the existence of the absolute space, with the rotation taken relative to the absolute frame creating the centrifugal force. Mach disagreed with the latter and its absolute space and argued that it is impossible to believe that the two phenomena are not related. The

13

important point is that the rotating stars could determine the motion of the arms. It must be possible to develop a more correct dynamics which ought to be able to explain how the rotating stars exert an "inertial force" on our arms.

However, Mach did not describe quantitatively how the distant stars and galaxies could exert a physical influence on laboratory phenomena. It is clear that this is a problem of gravitation. It follows that by criticizing the concept of an absolute space in discussing the anomaly of the centennial precession of Mercury's orbit and the inadequacy of special relativity, all ideas converge with gravitation.

1-6   Principle of Equivalence

As mentioned in section 1-3, the most important characteristic of gravity is the indistinguishability of gravitational mass from inertial mass. Before developing this viewpoint, it may be suitable to say a few words about an interesting characteristic of physical theories which is related to its methodology. We know that to make an affirmative assessment of a proposal (to prove it is true) is not easy, but to make a negative one is sometimes even more difficult. For example, to determine if "this matter can be done", it is enough that the matter has been done once or under only one situation. However, to show that "this matter cannot be done", it must be proved that the matter has never been accomplished nor can it ever be in any way. It is obviously impossible to prove this with experiments, since there is a limit to the number of experiments and one cannot perform an infinite number of experiments. Therefore, a negative assessment is always obtained by scientific abstraction from finite experiments.

This abstraction often brings about very significant consequences. It is possible to say that many fundamental principles are based on negative assessments. The most typical of these fundamental principles are the two laws of thermodynamics, which can be expressed as the impossibility of first and second kinds of machines in perpetual motion. One of the fundamental principles called the uncertainty relation in

14

quantum mechanics is also a negative thesis.

The relativity principle in special relativity is drawn from the impossibility of measuring absolute velocity. Einstein also used the same method for establishing a new theory, general relativity, when he considered that the first fundamental principle of gravitation could be expressed as a principle of "impossibility", i.e., that we cannot distinguish a uniform gravitational field from an accelerated frame. More specifically, within the limits of the locality of a frame, we cannot determine whether there is a gravitational field, as explained in the following discussion.

Einstein devised an ideal experiment to analyze the problem. He considered a lift-cabin which is equipped with various sorts of experimental instruments. When the lift-cabin is at rest with respect to the earth, the experimenter can see that all the objects inside are subjected to a force. If there is no other force to balance it, it will pull all these objects onto the floor of the cabin. According to these phenomena, the experimenter may readily conclude that his lift is being subjected to an external force that attracts objects to the floor. Now let the lift be in free-falling motion. Then, the experimenter will discover that the objects in the cabin are no longer subjected to the force as before and all objects have lost their previous acceleration. In other words, the objects in the cabin no longer show any sign of being subjected to the gravitational force; that is to say, the experimenter in observing the mechanical phenomena of any object cannot detect the slightest trace of a gravitational force.

Following this, Einstein made a further deduction that the experimenter in the cabin not only fails to detect any trace of gravitation by observing all mechanical phenomena, but also, as he affirms, in all other kinds of physical experiments. This means that in the frame of the lift, gravitation has been completely nullified. The lift experimenter cannot decide, by observing physical phenomena within the cabin, whether there exists an earth outside acting as a source of a gravitational force, nor can he measure whether his lift has an acceleration or not, just as in Salviati's ship the observer cannot make out if his ship is in motion or at rest.

In brief, in any local region we can find a frame of reference within which all the effects of the gravitational force may be completely eliminated, a fact which is the most important property of gravitation. In physics, no other force has this property. For example, neither the macroscopic electromagnetic force nor the strong and weak interactions in the realm of nuclei and particles can be completely eliminated through the choice of a suitable frame of reference.

The property of gravitation which enables gravity to be eliminated locally is commonly known as the principle of equivalence. This principle is the most fundamental characteristic of gravity.

Why must we emphasize "local limits" time and again? This is because real gravitational fields are only constant locally: gravitational acceleration g points to the earth's centre so that its direction varies from point to point around the earth, and its magnitude also varies, with height above the earth's surface. For a rigid lab falling freely, only its centre of mass is in free-fall, and all other parts of the lab, strictly speaking, are not in free-fall. Now, if there are two particles in the laboratory (see Fig. 1.4), their accelerations

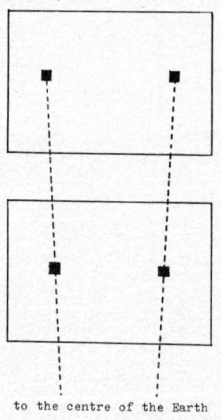

to the centre of the Earth

Fig. 1.4 Gravity can be eliminated
only for a local system falling freely

16

point towards the earth's centre and are not parallel to each other. They have a relative acceleration, which causes them to approach one another. Suppose the two particles are initially at rest relative to each other. It is easy to show that the relative speed with which the particles approach one another is given by

$$u = \frac{d}{R_\oplus^2} \sqrt{2G M_\oplus h} \quad \text{or} \quad u = \frac{d}{R_\oplus} \sqrt{2gh} \ ,$$

where h is the distance fallen, d their separation, $R_\oplus$ and $M_\oplus$ are the radius and mass of the earth respectively, and gravitational acceleration $g = GM_\oplus / R_\oplus^2$. The speed u is very small. For example, by taking $d = 10$ m, $h = 10$ km, then $u \approx 10^{-3}$ m/s.

The tides are brought about by the same relative acceleration mentioned above, so that the related force is called a tidal force. Since the tidal force cannot be eliminated by transforming the coordinates, it is possible, in principle, to detect the existence of gravitational fields. As u is proportional to d, the smaller the dimensions of a laboratory the more difficult it would be to carry out these observations. Strictly speaking, to state that all phenomena of gravity can be eliminated would be correct only for a point-like system in free-fall. This is the reason why we must emphasize the word "local".

1-7 Local Inertial Frames

For any local region in free fall, the principle of equivalence ensures that there exists a frame in which all effects of gravity are eliminated, and all particles without external forces (i.e. forces other than gravitational forces) acting on them are in rectilinear motion. They are called local inertial frames and conform to the definition of an inertial frame.

In general, in Newtonian mechanics, one can assess whether a frame is inertial by using the law of inertia: a body must be in inertial motion if no external force acts upon it. However, there can never actually be a laboratory "without any external forces", and inertial frames based on these abstractions have no practical value, because the

ubiquitous gravity cannot be screened. Only for local inertial frames
within which the law of inertia does indeed hold, can we really find
"the environment without any external forces". Therefore, the local
inertial frames are closer to the original meaning of an inertial frame.
Within local limits in spacetime, there is an infinite number of local
inertial frames, each moving with constant velocity relative to another.
Special relativity applies rigorously within these frames.

Local inertial frames are more restricted and more general than
Newton's inertial frames. First of all, the inhomogeneity of real
gravitational fields makes them only applicable locally instead of infinite in extent. Secondly, it is now obvious to us that the reason
why inertial frames are in an "advantageous position" is because of the
elimination of the effects of gravity. This point cannot be understood
in Newton's system. Furthermore, inertial frames in the Newtonian
system are given by the absolute space but not by the motion of the
matter, whereas, any laboratory in free-fall is a local inertial frame
and is influenced by the distribution and motion of matter. Now, local
inertial frames have to be unaccelerated relative to neither absolute
space nor one another (in the latter case, except within local limits).
For example, an artificial satellite rotating around the earth and a
spaceship flying under gravity to Venus, though they may be accelerated
relative each other, are both local inertial frames, because they move
only under the effects of gravity.

It is impossible, as mentioned in section 1-6, to eliminate all
effects of gravity from large regions. For instance, relative acceleration between two bodies cannot be eliminated by choosing a suitable
coordinate system. The acceleration expresses the differences in the
effects of gravity on these two points of spacetime and the relation
between the two local inertial frames.

Hence, the effects of gravity have precluded the existence of
inertial frames in a large region, and only local inertial frames,
between which relationships are determined by gravity, are possible.
In short, the effects of gravity are only in the determination of the

18

local inertial frames. The frames depend on gravity and the frames describe the spacetime background of the motion of the matter. Therefore, differing from other kinds of forces, gravity which influences the motion of the matter by determining the properties of spacetime, is itself described by the metric of spacetime.

1-8 Metrics of Spacetime

Consider a particle moving in spacetime. The motion consists of successive events, each of which is a point in spacetime, specified by four coordinates x^μ, where $\mu = 0,1,2,3$. It is conventional to take x^0 as a time coordinate t, and x^1, x^2 and x^3 as the three spatial coordinates.

In Newtonian mechanics the motion of the particle is described by $r(t)$, and as an essential quantity to describe the particle's motion, t is an absolute parameter and is independent of the choice of coordinates.

However, in relativity, t is no longer an absolute parameter and instead of t, we use another coordinate τ, the readings of a clock carried by the particle. The time τ is called the proper time. Obviously, τ is an invariant parameter (it has been introduced without mentioning any coordinate system). The motion of the particle is described by four functions $x^\mu(\tau)$, which describe what is known as the world line of the particle.

It may be useful to observe a simple analogy. In Euclidean geometry, a point is described by three coordinates x^i ($i = 1,2,3$), and curves are parametrised by using the arc length s, that is, they are specified by the three functions $x^i(s)$. The arc length is an invariant under coordinate transformations. Taking Cartesian coordinates x,y,z, the arc length between neighbouring points (x,y,z) and $(x+dx, y+dy, z+dz)$ is

$$ds^2 = dx^2 + dy^2 + dz^2 \ . \tag{1.6}$$

The same arc length in polar coordinates is

19

$$ds^2 = dr^2 + r^2 (d\theta^2 + \sin^2\theta \, d\phi^2) \ .$$

The analogous formula in spacetime gives $d\tau^2$ as a function of the coordinate differences dx^μ between any two neighbouring events. By choosing a freely falling frame at $x^\mu + dx^\mu$, i.e. a local inertial frame near the point x^μ, then according to the principle of equivalence, special relativity holds near this point and $d\tau^2$ can be expressed by Minkowski's formula, that is

$$d\tau^2 = dt^2 - \frac{1}{c^2} (dx^2 + dy^2 + dz^2) \tag{1.7}$$

where x, y, z, are still Cartesian spatial coordinates.

In ordinary space, we can use more general coordinates x^1, x^2, x^3, which are related to Cartesian coordinates by relations

$$x = x(x^1, x^2, x^3)$$
$$y = y(x^1, x^2, x^3)$$
$$z = z(x^1, x^2, x^3) \ ,$$

so that we have

$$dx = \frac{\partial x}{\partial x^1} dx^1 + \frac{\partial x}{\partial x^2} dx^2 + \frac{\partial x}{\partial x^3} dx^3$$

$$dy = \frac{\partial y}{\partial x^1} dx^1 + \frac{\partial y}{\partial x^2} dx^2 + \frac{\partial y}{\partial x^3} dx^3$$

$$dz = \frac{\partial z}{\partial x^1} dx^1 + \frac{\partial z}{\partial x^2} dx^2 + \frac{\partial z}{\partial x^3} dx^3 \ .$$

Thus

$$ds^2 = \left[(\frac{\partial x}{\partial x^1})^2 + (\frac{\partial y}{\partial x^1})^2 + (\frac{\partial z}{\partial x^1})^2 \right] (dx^1)^2$$
$$+ 2 \left[\frac{\partial x}{\partial x^1} \frac{\partial x}{\partial x^2} + \frac{\partial y}{\partial x^1} \frac{\partial y}{\partial x^2} + \frac{\partial z}{\partial x^1} \frac{\partial z}{\partial x^2} \right] dx^1 dx^2 + \dots \dots$$
$$= g_{ij} \, dx^i \, dx^j \ . \tag{1.8}$$

In the above equation, we have used Einstein's summation convention:

we shall sum over an index i for 1,2,3, if the index appears twice, once as an upper and once as a lower index. The coefficients g_{ij} in (1.8) are defined as

$$g_{ij} \equiv \frac{\partial x}{\partial x^i}\frac{\partial x}{\partial x^j} + \frac{\partial y}{\partial x^i}\frac{\partial y}{\partial x^j} + \frac{\partial z}{\partial x^i}\frac{\partial z}{\partial x^j} \quad.$$

The meaning of the formula (1.8) is quite simple: the distance ds is invariant, but the coordinate differences are not; they depend on the choice of coordinates. The functions $g_{ij}(x^1, x^2, x^3)$ involve nine components and can be written formally in the form of a matrix:

$$g_{ij} = \begin{pmatrix} g_{11} & g_{12} & g_{13} \\ g_{21} & g_{22} & g_{23} \\ g_{31} & g_{32} & g_{33} \end{pmatrix} \quad.$$

From the definition of g_{ij}, it can be deduced that $g_{ij} = g_{ji}$, i.e. only six components are independent. The set of functions g_{ij} is called the metric tensor, which gives the measuring characteristics of of the space. If Cartesian coordinates are used in Euclidean space, the simplest form of the metric tensor is obtained, occurring as

$$\delta_{ij} = \begin{pmatrix} 1 & 0 & 0 \\ 0 & 1 & 0 \\ 0 & 0 & 1 \end{pmatrix} \quad.$$

All of these concepts can be extended from three-dimensional space to four-dimensional spacetime without any difficulty. Using four-dimensional coordinates x^μ for describing the events and the world-line in spacetime, the separation of proper time (or in short, separation) between two events x^μ and $x^\mu + dx^\mu$ is

$$d\tau^2 = - g_{\mu\nu}dx^\mu dx^\nu \quad. \tag{1.9}$$

For different coordinate systems, the dx^μ may not be the same, but the separation $d\tau^2$ remains unchanged. The metric tensor $g_{\mu\nu}$ determines the geometric character of spacetime and has 4 x 4 = 16 components, of which only ten are independent, since by definition $g_{\mu\nu}$ is symmetric, i.e. $g_{\mu\nu} = g_{\nu\mu}$.

If we take a free-falling frame at a point in spacetime and use Minkowski's expression (1.7), then $g_{\mu\nu}$ become

$$\eta_{\mu\nu} = \begin{pmatrix} -1 & 0 & 0 & 0 \\ 0 & \dfrac{1}{c^2} & 0 & 0 \\ 0 & 0 & \dfrac{1}{c^2} & 0 \\ 0 & 0 & 0 & \dfrac{1}{c^2} \end{pmatrix},$$

which in (1.9) gives $d\tau^2 = -\eta_{\mu\nu}dx^{\mu}dx^{\nu}$.

We shall now point out some important differences between physical four-dimensional spacetime and Euclidean three-dimensional space.

Firstly, in Euclidean three-dimensional space, g_{ij} can always be reduced to δ_{ij} in the whole space by using an appropriate coordinate transformation. However, in spacetime it is generally possible, but only locally, to reduce $g_{\mu\nu}$ to $\eta_{\mu\nu}$. This is merely a reflection of the fact that in a gravitational field it is not possible to cover the whole of spacetime with a single inertial frame.

Secondly, in Euclidean three-dimensional space, the distance between different points in space must be larger than zero. However, in spacetime it is possible to find non-coincident events with positive, or negative, or zero separation. This is well-known in special relativity and can also be seen from the form of $\eta_{\mu\nu}$: there are three positive and one negative diagonal element in $\eta_{\mu\nu}$, so that $d\tau^2$ may have 2 possible signs. The difference between the number of positive and the number of negative diagonal elements is called the signature of the metric, which in this case is 2.

The separation which makes $d\tau^2 > 0$ is said to be time-like, expressing the possibility that the two events may be connected by a real physical signal. Consequently, there may be a causal connection between such events. The separation making $d\tau^2 < 0$ is said to be space-like, and the two events cannot be connected by any real physical process. Thus, there cannot be a causal connection between such events. Finally, $d\tau^2 = 0$ is said to be light-like, and only a body which moves

22

with the velocity of light can satisfy this condition.

Now that the determination of physical spacetime is the essence of gravity, the measure $g_{\mu\nu}$ of physical spacetime should be the essential quantity for describing gravity.

1-9 Geodesics

When a particle is in spacetime, which is defined by a metric tensor $g_{\mu\nu}$, how is its world line $x^{\mu}(\tau)$ to be determined? If no external forces were to act on the particle it would move in uniform rectilinear motion, that is, the following equation would be satisfied:

$$\frac{d^2 t}{d\tau^2} = 0 \quad , \qquad \frac{d^2 r}{d\tau^2} = 0 \quad .$$

However, under the condition that gravitational forces are present, there is no single inertial frame applicable for the whole spacetime, and the above forms cannot be used immediately.

Now, we will try to extend the equations mentioned above to circumstances with gravitational forces. In Euclidean three-dimensional space, a straight line corresponds to the shortest distance between two points in space. An analogous argument in the four-dimensional spacetime of special relativity would show that a straight line joining two time-like events is the longest. This means that if two clocks 1 and 2 set out, initially synchronised, from an event A and travel along different world lines (Fig. 1.5), they do not, in general, have the same time when they meet again at another event B (Fig. 1.5), and the proper time for rectilinear motion along the straight line joining A and B is always larger than that for a curved line. This has been named the "twin paradox" and has already been demonstrated by experiments.

23

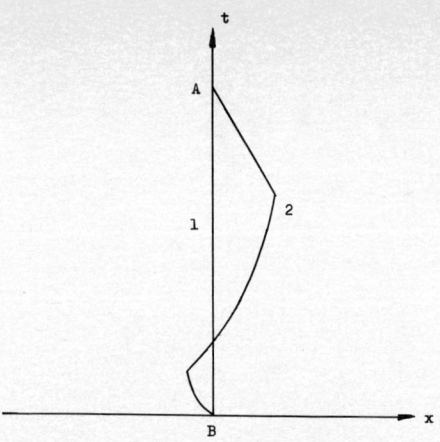

Fig. 1.5 The "Twin Paradox"

For example, by putting muons with 3.1 GeV energy in a storage
ring as has been done at CERN, it is possible to make two muons travel
along different world lines with the same starting event to meet at a
final event: with one muon at rest and the other in motion in a circular
orbit. The result is that the lifetime of the muon in motion is 64 microseconds, while the other one at rest has 2.2 microseconds. The
muons with different velocities have different separations of proper
time.

The separation between events A and B, along a world line
$x^\eta(\tau)$, can be written formally as

$$\tau_{AB} = \int_A^B \sqrt{- g_{\mu\nu} dx^\mu dx^\nu}$$
$$= \int_A^B \sqrt{- g_{\mu\nu}(x^\eta(\tau)) \frac{dx^\mu(\tau)}{d\tau} \frac{dx^\nu(\tau)}{d\tau} d\tau^2}$$

where η is a "floating" or unsummed index. If $x^\eta(\tau)$ is a straight
line, the value τ_{AB} is an extremum.

We can extend the above-mentioned result to general spaces, by
first introducing the concept of geodesics. In general, for spaces
defined by metric tensors $g_{\mu\nu}$, the extremal paths are known as geodesics. In spacetime, the metric tensors $g_{\mu\nu}$ are such that geodesics

24

along which the separation is zero may exist; these geodesics are called null geodesics. The above definition of geodesics is independent of the coordinate system.

In establishing general relativity, Einstein postulated that in a spacetime defined by a metric tensor $g_{\mu\nu}$, the world lines of material bodies not subjected to non-gravitational forces must be time-like geodesics. This is a generalization of the law which states that in inertial frames, bodies not subjected to any force must move along straight lines. The geodesic law of motion is very important in general relativity which has been formulated so that inertial and gravitational masses do not appear, representing concepts which differ very fundamentally from those of Newtonian gravitation.

It is worthwhile emphasising that the geodesic law applies to world lines between events in spacetime, and not to paths between positions in ordinary three-dimensional space, because any two positions in three-dimensional space may be connected not only by one path, but by a number of different paths. For example, as shown in Fig. 1.6a, under the action of gravity there are various possible parabolic motions from A to B. However, the world line of free motion between two events in spacetime is unique. The parabolas that begin at one point A and end at the same point B in Fig. 1.6a, do not pass through the same terminal point of event in four-dimensional spacetime (see Fig. 1.6b).

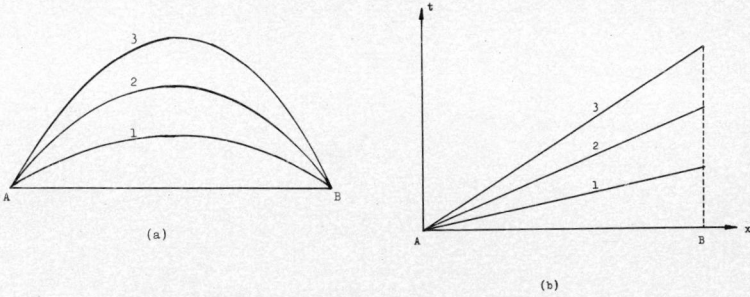

Fig. 1.6 Under the action of gravity, bodies may move in space along various paths. However, in spacetime graph (b), the world line of free motion between any two events is unique.

25

The equations satisfied by geodesics are:

$$\frac{d^2x^\mu}{d\tau^2} + \Gamma^\mu_{\nu\eta} \frac{dx^\nu}{d\tau} \frac{dx^\eta}{d\tau} = 0 \quad , \qquad (1.10)$$

where $\Gamma^\mu_{\nu\mu}$ are called Christoffel symbols which are defined by $g_{\mu\nu}$

$$\Gamma^\mu_{\nu\eta} = \frac{1}{2} g^{\mu\alpha} \left\{ \frac{\partial g_{\alpha\eta}}{\partial x^\nu} + \frac{\partial g_{\alpha\nu}}{\partial x^\eta} - \frac{\partial g_{\nu\eta}}{\partial x^\alpha} \right\} \quad ,$$

where $g^{\mu\alpha}$ is the inverse of $g_{\mu\alpha}$, that is, $g^{\mu\alpha}g_{\alpha\nu} = \delta^\mu_\nu$; and δ^μ_ν is the Kronecker symbol defined as

$$\delta^\mu_\nu = \begin{cases} 1 & \mu = \nu \\ 0 & \text{otherwise} \end{cases} \quad . \qquad (1.11)$$

1-10 Curved Spaces

Let us now consider a question: what is the geometrical meaning of the assertion that only locally inertial frames exist?

If there were just one inertial frame for the whole of space, we would be able to find a coordinate system in which all free motion could become uniform rectilinear ones. An example is that various parabolic paths as seen near the Earth's surface may appear straight if we observe in the frame of a lift in free-fall, but this is impossible under the action of gravity in a different frame such as that at rest with respect to the Earth. In other words, it is not possible in the presence of gravity, and hence curvature, to transform all geodesics into straight lines, that is, under such circumstances geodesics are intrinsically curved (because of the characteristics of spacetime). As a result, spacetime itself is said to be curved.

To explain the concept of curvature of spacetime, we would like to consider as a start the following two-dimensional surfaces: a plane, the surface of a sphere and that of a cylinder. It seems possible to determine whether these surfaces are flat or curved. The plane is obviously flat, the sphere is curved and cannot be made into a plane by shearing and stretching, and as for the cylinder, it can be unrolled into a plane.

It is easy to discuss two-dimensional surfaces since we live in a three-dimensional space. Consider a further question: can we study two-dimensional surfaces without observations in a three-dimensional space? Imagine a being which lives and measures only in a two-dimensional surface, and it is not permitted to travel at will in a three-dimensional space to make observations. Can this being determine whether the surface in which it finds itself is curved or flat? This is not an easy question to reply to at once. "One cannot be sure of the true sights of the Lu mountain, since one is on it." Gauss, however, has given us the answer.

Euclidean plane geometry asserts that in a plane, the angles of a triangle add up to 180°. On the surface of a sphere, however, the sum of the angles of a "geodesic triangle" always exceed 180° (Fig. 1.7). Therefore the two-dimensional being can determine whether the surface on which it lives is curved by measuring the angle sum of a "geodesic triangle" on the surface. Is the three-dimensional space in which we live curved or flat? This can also be resolved analogously. Gauss himself measured the angle sum of a triangle formed by three mountains as vertices, but failed to detect any departure from 180° within the limits of accuracy of his experiments.

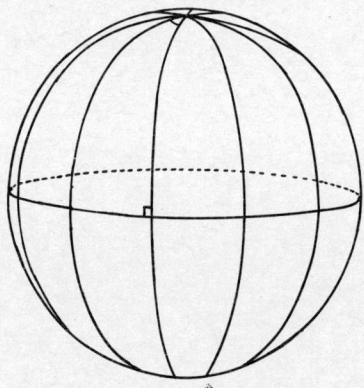

Fig. 1.7 Angle sum of spherical triangle exceeds 180°

Let us consider a circle on a curved surface. To define a circle of "radius" a and centre O on an arbitrary surface, we draw all geodesics emanating from O, and mark on each geodesic the point whose distance along the geodesic from O is a, the locus of all these points being the required circle of "radius" a. For instance, let us try this on a sphere of radius R (Fig. 1.8). The circumference of the circle is

$$c = 2\pi R \sin \frac{a}{R} = 2\pi a (1 - \frac{a^2}{6R^2} + ...) \quad .$$

We define the curvature K of the sphere as

$$K \equiv \frac{3}{\pi} \lim_{a \to 0} (\frac{2\pi a - c}{a^3}) \quad . \tag{1.12}$$

Then using the formula for the circle's circumference, we have

$$K = \frac{1}{R^2} \quad .$$

This is a very useful result. It tells us that we can obtain the curvature of a sphere from measurements taken on the sphere's surface (a and c are all internal quantities of the sphere's surface). The same procedure can be extended to more general cases; that is, we can define the curvature of an arbitrary surface.

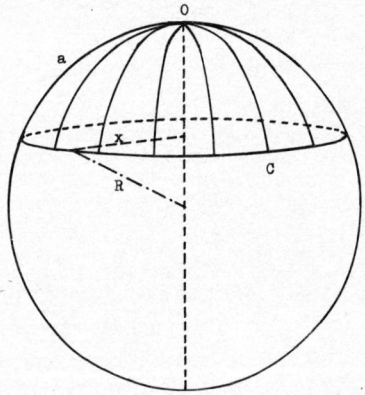

Fig. 1.8 Circle of "radius" a on sphere of radius R

Curvature may be positive or negative. If the circumference c exceeds $2\pi a$, the corresponding surface has a negative curvature. This is illustrated in Fig. 1.9, which shows three surfaces with curvatures which are zero, positive and negative.

Now, we use the metric tensor g_{ij} of two-dimensional surfaces to express various properties discussed above. In a plane, if we use Cartesian coordinates, $x^1 = x$, $x^2 = y$, then

$$ds^2 = dx^2 + dy^2 = (dx^1)^2 + (dx^2)^2 \ , \qquad (1.13)$$

that is

$$g_{ij} = \begin{pmatrix} 1 & 0 \\ 0 & 1 \end{pmatrix} \ . \qquad (1.14)$$

In these coordinates, the relations of plane geometry are obviously satisfied, indicating we are on a flat surface. However, we could equally well have used polar coordinates, $x^1 = r$, $x^2 = \phi$. Then

$$ds^2 = dr^2 + r^2 d\phi^2 = (dx^1)^2 + (x^1)^2 (dx^2)^2 \ , \qquad (1.15)$$

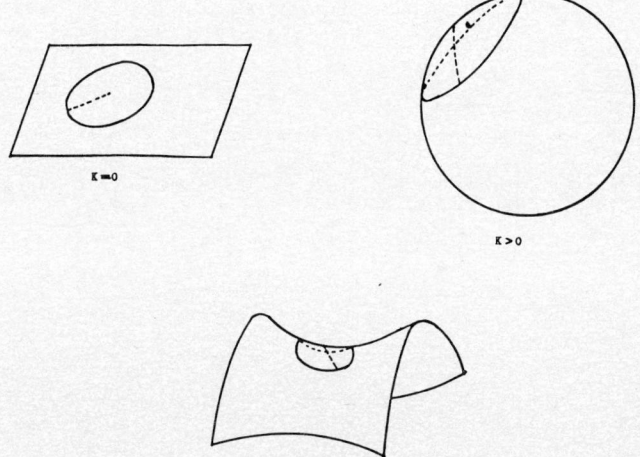

Fig. 1.9 Three kinds of surfaces with different curvatures.

29

that is,

$$g_{ij} = \begin{pmatrix} 1 & 0 \\ 0 & (x^1)^2 \end{pmatrix} .$$

Although the form of this metric differs from that in Cartesian coordinates, it still describes the same flat surface. Hence, we should define a flat surface as follows: if we can find a coordinate transformation which can change the form of g_{ij} into that of (1.14), then the two-dimensional surface is intrinsically flat. The transformation between polar coordinates (r, ϕ) and Cartesian coordinates (x,y) does indeed exist, as

$$x = r \cos \phi ,$$
$$y = r \sin \phi . \tag{1.16}$$

Let us analyse a cylinder of radius R (Fig. 1.10). On the apparently curved surface, the formula for a small distance on the surface can be written as:

$$ds^2 = dz^2 + R^2 d\phi^2 . \tag{1.17}$$

It is easy to show that this is flat since by defining coordinates

$$x^1 = z, \quad x^2 = R\phi ,$$

Equation (1.17) becomes

$$ds^2 = (dx^1)^2 + (dx^2)^2 ,$$

which is the same as the line element on a plane. It must be noted that a cylinder is not the same as a plane in all respects. For example, among all the geodesics through any point on a cylinder, one will be closed (see Fig. 1.10), which does not happen on a plane. Differences of this sort are actually differences in the connectivity of the surface as a whole, or topological differences. Curvature, on the other hand, is a local property and, according to the definition (1.12), can be established by measuring small distances.

30

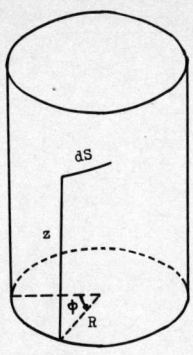

Fig. 1.10 Line element on a cylinder

As for the surface of a sphere of radius R, we can use polar coordinates θ and ϕ to describe every point (Fig. 1.11). Taking $x^1 = \theta$, $x^2 = \phi$, the line element is

$$ds^2 = R^2 d\theta^2 + R^2 \sin^2\theta \, d\phi^2$$
$$= R^2(dx^1)^2 + R^2 \sin^2 x^1 \, (dx^2)^2 \quad , \tag{1.18}$$

and the metric tensor is

$$g_{ij} = \begin{pmatrix} R^2 & 0 \\ 0 & R^2\sin^2 x^1 \end{pmatrix} \quad . \tag{1.19}$$

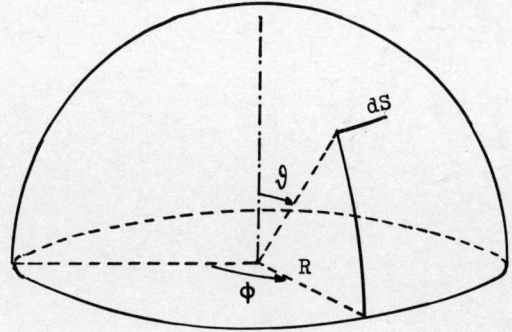

Fig. 1.11 Line element on a sphere

31

This time, we cannot find new coordinates $x^{1'} = x^{1'}(\theta, \phi)$, $x^{2'} = x^{2'}(\theta, \phi)$, which can transform (1.19) into (1.14). Therefore, a spherical surface is intrinsically curved.

In general, how can we be certain that there exists a transformation which can reduce g_{ij} into (1.14)? Can we know whether the space concerned is curved just by considering its tensor g_{ij}? These questions can be answered by the curvature: if curvature $K \neq 0$, the surface is not flat; if $K = 0$, it is flat, and there must be a transformation which can transform g_{ij} into (1.14). Therefore, the problem has become one of how to calculate the curvature K from g_{ij}. Gauss has given a general formula for calculating K from g_{ij} as

$$K = \frac{g_{ij}}{g} R^j_{112}$$

where

$$R^j_{112} = \frac{\partial \Gamma^j_{11}}{\partial x^2} - \frac{\partial \Gamma^j_{12}}{\partial x^1} + \Gamma^i_{11} \Gamma^j_{2i} - \Gamma^i_{12} \Gamma^j_{1i} \quad ,$$

$$g = g_{11}g_{22} - g_{12}g_{21}.$$

Since the general formula is quite complicated, we also write Gauss's formula for a particular case $g_{12} = g_{21} = 0$, as follows:

$$K = - \frac{1}{2\sqrt{g}} \left\{ \frac{\partial}{\partial x^2} \left(\frac{1}{\sqrt{g}} \frac{\partial g_{11}}{\partial x^2} \right) + \frac{\partial}{\partial x^1} \left(\frac{1}{\sqrt{g}} \frac{\partial g_{22}}{\partial x^1} \right) \right\} \quad . \tag{1.20}$$

For a two-dimensional curved surface, it is always possible to transform the metric tensor g_{ij} to diagonal form, $g_{12} = g_{21} = 0$. This form of metric is called orthogonal. Substituting the polar coordinates of a plane into (1.20), we find $K = 0$. It is clear that K is invariant, that is, no matter what coordinate system we use, we always get the same value for K. It is therefore an intrinsic geometrical property of the surface itself. For example, the curvature K of a spherical surface is always equal to $1/R^2$, which cannot be made zero by transforming the coordinates.

The value of K itself, generally speaking, is a function of the

32

point chosen, that is, the curvature of general surfaces varies from point to point. For different points on a spherical surface, K is constant and equal to $1/R^2$. This is a particular case. In general, a surface whose curvature has the same value at all points is called a constant curvature surface, which includes the sphere, the cylinder and the plane.

1-11   Relation between Curvature and Matter

We will now extend the concepts that were introduced for two-dimensional space to physical four-dimensional spacetime. For spaces of more than two dimensions, it is not possible to specify curvature by just one function K. For example, we can draw a geodesic triangle in the neighbourhood of a point in space so that the departure of the angle-sum from 180^0 gives the unique curvature of the space (close to this point). However, in the neighbourhood of a spacetime point, there are many different possible geodesic triangles, each of which may give a different angle-sum departure from 180^0. It turns out that we must use not just one but a whole group of quantities to describe the curvature of spacetime. This is the so-called Riemannian metric tensor $R_{\mu\nu\alpha\eta}$, which is defined as

$$R_{\mu\nu\alpha\eta} \equiv g_{\mu\beta} R^{\beta}_{\nu\alpha\eta}$$

$$R^{\mu}_{\nu\alpha\eta} \equiv \frac{\partial \Gamma^{\mu}_{\nu\alpha}}{\partial x^{\eta}} - \frac{\partial \Gamma^{\mu}_{\nu\eta}}{\partial x^{\alpha}} + \Gamma^{\beta}_{\nu\alpha} \Gamma^{\mu}_{\eta\beta} - \Gamma^{\beta}_{\nu\eta} \Gamma^{\mu}_{\alpha\beta} \quad . \tag{1.21}$$

The tensor $R_{\mu\nu\alpha\eta}$ has the following symmetries:

$$R_{\mu\nu\alpha\eta} = R_{\alpha\eta\mu\nu}$$

$$R_{\mu\nu\alpha\eta} = -R_{\mu\nu\eta\alpha} = -R_{\nu\mu\alpha\eta} = R_{\nu\mu\eta\alpha}$$

$$R_{\mu\nu\alpha\eta} + R_{\mu\eta\nu\alpha} + R_{\mu\alpha\eta\nu} = 0 \quad . \tag{1.22}$$

Thus, Riemannian curvature has $\frac{1}{12} n^2(n^2 - 1)$ independent components in n dimensions. In two dimensions, $n = 2$, $\frac{1}{12} n^2(n^2 - 1) = 1$, that is, only one independent component is enough to describe it, as was seen in section 1-10. In physical spacetime, $n = 4$, and the number of independent components is $\frac{1}{12} 4^2(4^2 - 1) = 20$.

In astrophysics, we shall often deal with spaces and spacetimes with symmetries which will simplify the problem. Firstly, we consider a three-dimensional space with constant curvature. It was pointed out in section 1-10 that a two-dimensional spherical surface is a two-dimensional space with constant curvature. Analogous methods will be used for discussing three-dimensional space with constant curvature.

Choosing a point in Euclidean three-dimensional space as the centre of a spherical surface with r as the radius, the area of the surface is $4\pi r^2$. Using polar coordinates, with the origin at the centre of the sphere, positions on the surface are specified by θ and ϕ, and a small displacement on the surface can be expressed as

$$ds^2 = r^2 d\theta^2 + r^2 \sin^2\theta \, d\phi^2 \ .$$

For a three-dimensional space with constant curvature, we can use analogous spherical polar coordinates, taking a series of spheres whose centres coincide at the origin and radius r being constant for each sphere, with the area of a spherical surface with constant r determined as $4\pi r^2$. Thus we can still employ the coordinates θ and ϕ for the spherical surfaces. The difference is that the ratio of the area to the radius of the sphere is in general not equal to $4\pi r$, that is, if the surface area of the sphere is equal to $4\pi r^2$, the radius is generally not r. Therefore, the metric in this space in general is

$$ds^2 = f(r)dr^2 + r^2 d\theta^2 + r^2 \sin^2\theta \, d\phi^2$$

where $\sqrt{f(r)}dr$ is the proper distance between neighbouring points (r, θ, ϕ) and $(r+dr, \theta, \phi)$.

We demand that all geodesic surfaces in the space must have the same curvature K. For example, if we choose the equatorial surface $\theta = \pi/2$, then $d\theta = 0$ and

$$ds^2\Big|_{\theta=\pi/2} = f(r)dr^2 + r^2 d\phi^2 \ .$$

In this surface, taking $x^1 = r$, $x^2 = \phi$, we have

$$ds^2 = f(x^1)(dx^1)^2 + (x^1)^2(dx^2)^2 \ ,$$

that is,

$$g_{ij} = \begin{pmatrix} f(x^1) & 0 \\ 0 & (x^1)^2 \end{pmatrix}.$$

This is a two-dimensional surface, and its metric is orthogonal. Therefore, the curvature can be calculated by using Gauss's formula (1.20), which gives

$$K = \frac{\dfrac{df}{dx^1}}{2(f)^2 x^1}$$

or

$$\frac{\dfrac{df}{dx^1}}{(f)^2} = -\frac{d}{dx^1}\left(\frac{1}{f}\right) = 2Kx^1 .$$

where $f = f(x^1)$. If K is constant, the equation can be solved as follows:

$$\frac{1}{f(x^1)} = C - K(x^1)^2 ,$$

here C is a constant of integration. For flat space, $K = 0$, so that $f = 1/C$. If we demand $f(0) = 1$, then $C = 1$. The metric in this constant curvature space is therefore

$$ds^2 = \frac{dr^2}{1 - Kr^2} + r^2 d\theta^2 + r^2 \sin^2\theta d\phi^2 . \tag{1.23}$$

The proper radius of the r-sphere is

$$a(r) = \int_0^{a(r)} ds = \int_0^r \frac{dr}{\sqrt{1 - Kr^2}} = \frac{1}{\sqrt{K}} \sin^{-1}(r\sqrt{K}) ,$$

or

$$r = \frac{1}{\sqrt{K}} \sin(a\sqrt{K}) .$$

Now, the relationship between area $4\pi r^2$ and proper radius a is

$$A = 4\pi r^2 = \frac{4\pi}{K} \sin^2(a\sqrt{K}) . \tag{1.24}$$

When $a\sqrt{K} \ll 1$, the area A approximates to

$$A = 4\pi a^2 \quad ,$$

which is that in flat space. As a increases, A clearly departs from $4\pi a^2$ in a way which depends on the sign of K. If K < 0, we have

$$A = \frac{4\pi}{|K|} \sinh^2(a\sqrt{K}) \quad , \tag{1.25}$$

which shows that the area increases faster than in flat space. On the other hand, if K > 0, A increases more slowly than $4\pi a^2$, and reaches a maximum value

$$A_{max} = \frac{4\pi}{K}$$

when $a\sqrt{K} = \frac{\pi}{2}$. As the radius a increases further, A decreases and becomes zero when $a = \pi/\sqrt{K}$. This means that this positively-curved space is closed. The behaviour is closely analogous to that of the circumference of circles on a surface of constant curvature: as r increases, the length of the circumference increases first, and then reaches a maximum value (when $r = \frac{\pi}{2} R$); as r increases further, the circumference decreases, and becomes zero when $r = \pi R$.

Using this example, the meaning of the assertion that gravity determines the metrics of spacetime is clearer. First of all, when no gravitational field exists, the spacetime metric g_{ij} of special relativity can be applied to the whole of spacetime. It is impossible to cover spacetime with Minkowskian coordinates of special relativity if there is a non-homogeneous gravitational field present. This property is precisely analogous to curved surfaces. The metric properties on a sufficiently localised scale in curved surfaces are all analogous to the two-dimensional plane. For instance, as $a \rightarrow 0$, the length of the circumference $c \rightarrow 2\pi a$, which is the same as flat space. There is no such relation on a large scale, so that we can say that gravity makes spacetime curved.

The origin of gravity is matter. How does matter determine the curvature of spacetime? This is answered by Einstein's field equations. The definite form of the equations is as follows:

$$R_{\mu\nu} - \frac{1}{2} g_{\mu\nu} R = - \frac{8\pi G}{c^4} T_{\mu\nu} \qquad (1.26)$$

where $R_{\mu\nu} = g^{\alpha\beta} R_{\mu\alpha\nu\beta}$, $R = g^{\mu\nu} R_{\mu\nu}$, and $T_{\mu\nu}$ is the stress-energy-momentum tensor of matter. For example, for dust-like matter, we have

$$T^{\mu\nu} = \rho c^2 \frac{dx^\mu}{d\tau} \frac{dx^\nu}{d\tau} \qquad (1.27)$$

where ρ is the density of the dust in a coordinate system which moves with the dust. One more example is an ideal fluid for which

$$T^{\mu\nu} = pg^{\mu\nu} + (p + \rho c^2) \frac{dx^\mu}{d\tau} \frac{dx^\nu}{d\tau} , \qquad (1.28)$$

where p is pressure of the fluid.

The gravitational field equation (1.26) is equivalent to Poisson's equation

$$\nabla^2 \phi = - 4\pi \rho G \qquad (1.29)$$

in Newton's gravitation. In fact, (1.26) can indeed be transformed into (1.29) on the scale in which Newton's theory holds. However, it should be stressed that the fundamental starting points of (1.26) and (1.29) are completely different. On the left-hand side of Einstein's equation are the quantities which describe the properties of matter. Relation (1.26) tells us the effects of matter and its motion on space-time. It should be realised that the curvature of spacetime and the strength of the gravitational field are two different ways of describing the same thing. A large curvature of spacetime is equivalent to a strong gravitational field, and vice versa.

The fact that Newton's theory is valid on a very large scale shows that in such cases, all gravitational fields can be considered weak. How are the strengths of the fields to be assessed? Consider a system of mass M and dimension R, then the quantity which shows the strength of its gravity is

$$r_s = 2\,GM/c^2 \quad,$$

which is called the gravitational radius. If $r_s \ll R$, the field is weak and the curvature of spacetime is very small; if $r \lesssim R$, there is a strong field and the curvature is very large. The values of $r_s/R = 2\,GM/Rc^2$ for some typical celestial bodies are shown below in Table 1.

Table 1

Celestial body	$2GM/Rc^2$	Celestial body	$2GM/Rc^2$
Moon	$10^{-10.1}$	Neutron Star	10^{-1}
Earth	$10^{-8.9}$	Galaxy	10^{-6}
Sun	$10^{-5.4}$	Universe	1
White Dwarf	10^{-4}		

It can be seen that the majority of common celestial bodies have weak fields for which Newton's theory holds. It is not entirely surprising that Gauss's measurements failed to detect the curvature of space near the earth's surface.

REFERENCES

Textbooks on general relativity

L. Landau and E. Lifshitz, The Classical Theory of Fields (1962, Pergamon Press, Oxford).

R. Adler, M. Bazin and M. Schiffer, Introduction to General Relativity, 2nd Ed. (1975, McGraw-Hill, New York).

Basic theoretical works involving subjects on astrophysics

M. Berry, Principles of Cosmology and Gravitation (1976, Cambridge Univ. Press).

S. Weinberg, Gravitation and Cosmology (1972, John Wiley).

C.W. Misner, K.S. Thorne and J.A. Wheeler, Gravitation (1973, W.H. Freeman and Co.).

Ya. B. Zel'dovich and I.D. Novikov, Relativistic Astrophysics, Vol. 1
(1971, Univ. of Chicago Press, Chicago).

Chapter 2

EFFECTS IN A WEAK GRAVITATIONAL FIELD

2-1   Gravitational Redshift

A system for which $(2GM/Rc^2) \ll 1$, involves what is called a weak gravitational field. For such a system, Newtonian gravitation is sufficient and only very little correction need to be brought in using general relativity. However, for quite accurate measurements, such deviations are clearly observable. These weak field effects which will be discussed in this chapter, are of great importance in testing the validity of general relativity and its application to astrophysics.

It is possible to discuss gravitational effects by just using the principle of equivalence without recourse to Einstein's gravitational redshift, that is, how the frequency of light changes when it propagates in a gravitational field.

The energy of a photon of frequency ν is $h\nu$, and the inertial mass of the photon is then $h\nu/c^2$. According to the principle of equivalence, which is also obeyed by photons, the ratio of the inertial mass to the gravitational mass is always the same for any material body.

Suppose light propagates in the gravitational field near the Earth where the gravitational acceleration is g. Let us consider a laboratory at height h in free-fall and assume that a light ray with frequency ν_e moves from the floor through space to arrive and be detected at a later time on the ceiling.

According to the principle of equivalence, there is no gravitational force in this frame, so that the propagation of light will be described by special relativity with the speed of light c. Therefore, the propagation will take time t = h/c from the floor to the ceiling, and its frequency arriving at the ceiling will still be ν_e.

This view is not shared by an observer at rest relative to the Earth, to which the laboratory is in accelerated motion. During the time t, the downward velocity increases by u = gt = gh/c. If the laboratory is at rest relative to the Earth at the moment of emission, it will be moving downwards with velocity u at the moment of reception on the ceiling. Therefore, for an observer at rest relative to the Earth, the observer on the ceiling is not at rest but moves downwards with velocity u (Fig. 2.1). The photons as observed by the former will have a frequency less than ν_e. The decrease should be equal to that due to the Doppler effect at velocity u, that is, the frequency obtained by the observer at rest in the Earth's frame is

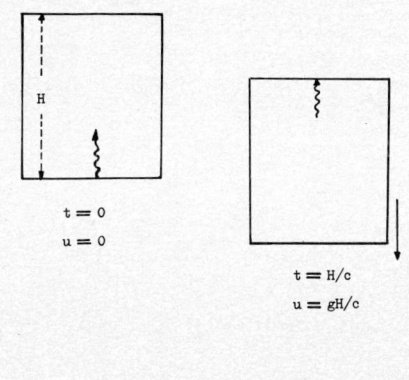

$$t = 0$$
$$u = 0$$

$$t = H/c$$
$$u = gH/c$$

Fig. 2.1 Gravitational redshift effect

41

$$\nu = \nu_e\left(1 + \frac{u}{c}\right)^{-1} = \nu_e\left(1 + g\,\frac{h}{c^2}\right)^{-1} \; .$$

Since $gh/c^2 \ll 1$ (weak gravitational field), the above formula can be written as

$$\frac{\nu - \nu_e}{\nu_e} \approx -\, g\,\frac{h}{c^2} \qquad\qquad (2.1)$$

where gh is the difference in the gravitational potential between the points of emission and reception. Thus, (2.1) can be rewritten in a more general form as

$$\frac{\nu_2 - \nu_1}{\nu_1} \approx \frac{1}{c^2}\,[\phi(x_1) - \phi(x_2)] \; , \qquad\qquad (2.2)$$

which states that when a light ray of frequency ν_1 emitted at x_1 arrives at x_2, its frequency becomes ν_2, where $\phi(x_1)$, and $\phi(x_2)$ denote the gravitational potentials at x_1 and x_2 respectively. Notice the condition under which (2.2) holds: observers at x_1 and x_2 are both at rest relative to the gravitational field.

By means of the Mössbauer effect, Pound, Rebka and Snider have carried out the verification of (2.2) in a terrestrial laboratory. They set the source ^{57}Co, which emits 14.4 keV gamma radiation, 22.6 m higher (or further away from the centre of the Earth) than the receiver ^{57}Fe. The violet shift would have an observed value of

$$\frac{\Delta\nu_G}{\nu_e} = \frac{\nu - \nu_e}{\nu_e} = 2.46 \times 10^{-15} \; .$$

This shift is very small and even less than the resonance-absorption width Γ of ^{57}Fe. In order to surmount the difficulty in measurement, the modulation method was used. Since the resonance-absorption cross-section of ^{57}Fe is proportional to

$$C(\Delta\nu) = \frac{\Gamma}{(\Delta\nu)^2 + \Gamma} \; ,$$

where $\dfrac{\Gamma}{\nu} \approx 1.13 \times 10^{-12}$,

if the source moves back and forth in simple harmonic motion with velocity $v = v_0 \cos\omega t$, then the frequency shift due to the Doppler effect would be

$$\frac{\Delta\nu_D}{\nu} = -\frac{v_0}{c}\cos\omega t \quad .$$

Now, $\Delta\nu$ in $C(\Delta\nu)$ is dependent on two factors: one is the gravitational red-shift (or violet-shift) $\Delta\nu_G$, and the other is the Doppler effect $\Delta\nu_D$ due to the motion of the source. Hence $\Delta\nu = \Delta\nu_G + \Delta\nu_D$, which gives

$$C(t) = \frac{\Gamma^2}{(\Delta\nu_G + \Delta\nu_D)^2 + \Gamma^2} = \frac{(\Gamma/\nu)^2}{(\Delta\nu_G/\nu + \Delta\nu_D/\nu)^2 + (\Gamma/\nu)^2}$$

$$\simeq \frac{(\Gamma/\nu)^2}{(v_0^2/c^2)\cos^2\omega t + (\Gamma/\nu)^2}\left\{1 - \frac{(2\Delta\nu_G \, v_0/\nu c)\cos\omega t}{v_0^2/c^2 \cos^2\omega t + (\Gamma/\nu)^2}\right\} \quad .$$

It follows that the linear term of $\cos\omega t$ will be dependent on $\Delta\nu_G$. By taking this term from the change in the resonance-absorption time, $\Delta\nu_G$ can be measured. Accordingly, the result obtained is

$$\frac{\Delta\nu_G}{\nu} \simeq (2.57 \pm 0.26) \times 10^{-15} \quad ,$$

which agrees quite well with the prediction given by (2.1).

It can be shown, by using (2.2), that the redshift of a light ray emitted from the surface of the Sun and measured in a terrestrial laboratory is

$$\frac{\Delta\nu}{\nu} \approx \frac{1}{c^2}\left(-\frac{GM_\odot}{R_\odot} + \frac{GM_\oplus}{R_\oplus}\right) \approx -\frac{1}{c^2}\frac{GM_\odot}{R_\odot} = -2 \times 10^{-6} \quad ,$$

and observations agree with this result to within 5% of allowable errors.

More recently, a hydrogen maser clock was flown in a rocket at an altitude of 10 000 km to test the redshift prediction. The first order Doppler shift of up to 2×10^{-5} had to be determined or eliminated

43

with sufficient accuracy in order to observe the predicted gravitational effect of at most 4×10^{-10}. A clever arrangement of the tracking instrumentation provided the required cancellation of the first-
order Doppler shift, except for that portion which resulted from the
acceleration of the tracking station during the round-trip propagation
of the signals to the spacecraft. This was to be accounted for in the
post-flight analysis. Similarly, the effect of the neutral atmosphere
on the signal propagation was cancelled to the extent that the atmosphe-
ric contribution was constant during the 0.1 s round-trip propagation
time from ground to rocket. A judicious choice of tracking frequencies
and technique also resulted in the virtual cancellation of the contribution of the ionosphere. The predicted gravitational redshift was
confirmed to well within the uncertainty of 140 parts per million.

The physical meaning of gravitational redshift can be understood
in another context: if an atom emitting light is regarded as a "clock",
and a process of emission of an integral wave corresponds to an interval, then for an observer at infinity, the frequency of light emitted
by an atom in a gravitational field becomes lower, which is equivalent
to a retardation of the clock. If $\Delta t(r)$ denotes the period of an
integral wave emitted on the surface of a star of mass M and radius
r, then $\Delta t(\infty)$ denotes that of an identical atom at infinity, and we
have the ratio:

$$\frac{\Delta t(r)}{\Delta t(\infty)} = 1 + \frac{GM}{rc^2} \ .$$

2-2 Schwarzschild Metric

In order to discuss strictly the various effects caused by
gravitation, it is necessary to know the solution to Einstein's field
equations, i.e. the spacetime metric $g_{\mu\nu}(x)$.

Schwarzschild derived a rigorous solution of Einstein's field
equations in 1916, known as the 'Schwarzschild metric', which is
applied generally to astrophysics. This metric expresses the spacetime
curvature near a spherically symmetric body of mass M. We shall not

44

derive the metric in detail, but will only discuss the physical meaning of the solution.

For a spherically symmetric system of mass M, we can choose polar coordinates with the origin at O. Thus a family of curved surfaces, each with constant r, is a series of concentric spheres on which it is most natural to adopt the coordinate r so that a sphere with constant r has area $4\pi r^2$, and the metric on the surface of the sphere would then be

$$ds^2 = r^2d\theta^2 + r^2\sin^2\theta d\phi^2 \quad .$$

Such a definition of r, as noted in the preceding chapter, is no longer the distance from the origin to the surface, because of the curvature of space caused by mass M. Therefore, the spatial metric should be expressed as

$$ds^2 = f(r)dr^2 + r^2d\theta^2 + r^2\sin^2\theta d\phi^2 \quad .$$

In addition, it is now known, from the previous discussion of gravitational redshift, that the existence of the mass M can also cause the clock on each r-sphere to run so that it is no longer observed from other r-spheres to be at the same rate. To allow for these effects, we write the general form of a metric near a spherically symmetric mass M as

$$d\tau^2 = e(r)dt^2 - \frac{1}{c^2}[f(r)dr^2 + r^2d\theta^2 + r^2\sin^2\theta d\phi^2] \quad .$$

In the case where M = 0, space would be flat, that is

$$e(r) = f(r) = 1 \quad .$$

Mass M causes e and f to depart from unity. By rigorously solving the Einstein equations (1.24), these functions can be found to be

$$e(r) = f^{-1}(r) = 1 - \frac{2GM}{c^2r} \quad ,$$

i.e. the Schwarzschild metric has the following form:

45

$$d\tau^2 = \left(1 - \frac{2GM}{c^2r}\right) dt^2$$

$$- \frac{1}{c^2}\left(\frac{dr^2}{1 - 2GM/c^2r} + r^2 d\theta^2 + r^2 \sin^2\theta d\phi^2\right) \quad , \qquad (2.3)$$

namely

$$g_{\mu\nu} = \begin{pmatrix} -\left(1 - \frac{2GM}{c^2r}\right) & 0 & 0 & 0 \\ 0 & \frac{1}{c^2}\left(1 - \frac{2GM}{c^2r}\right)^{-1} & 0 & 0 \\ 0 & 0 & \frac{1}{c^2}r^2 & 0 \\ 0 & 0 & 0 & \frac{1}{c^2}r^2\sin^2\theta \end{pmatrix} .$$

This is the well known expression for the "Schwarzschild metric", valid in the exterior region of spherical celestial bodies, such as the Sun and the Earth. We shall test this point in the next section.

2-3   Clocks Moving Around the Earth

In the preceding chapter we discussed how to measure the interval of proper time between two events, A and B, with time-like separation. Choose a world line of the measuring clock through the points A and B, the initial and final events, where the time spent by the clock between A and B is a proper separation. It has also been pointed out that the consequences of measurement, generally speaking, depend on the world line chosen. Different world lines between A and B correspond to different proper times.

For instance, consider two identical clocks 1 and 2, initially synchronised with the same time, at rest on the Earth's surface. Let 1 remain on the ground while 2 is flown around the Earth at height h with speed v relative to the ground (Fig. 2.2). After one circumnavigation, 1 and 2 are compared, that is, the proper time τ_1 experienced by 1 is compared with the proper time τ_2 experienced by 2.

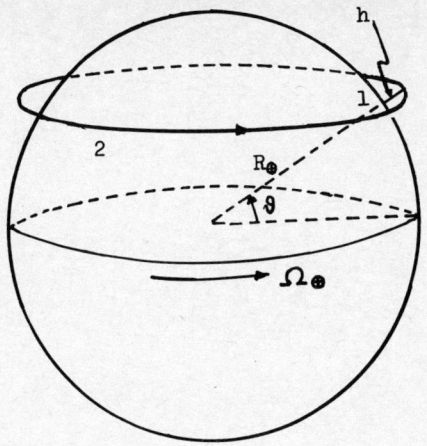

Fig. 2.2 Clocks moving around the Earth

We do not expect τ_1 and τ_2 to be the same, because 1 and 2 have pursued different world lines although they may have identical initial and final events.

We shall use the Schwarzschild metric* to calculate the expressions for τ_1 and τ_2. For 1, we have $r = R$, $dr = 0$ and $d\theta = 0$. Thus,

$$\tau_1 = \int \sqrt{(1 - 2GM_\oplus /c^2 R_\oplus)dt^2 - R_\oplus^2 \cos^2\theta d\phi^2/c^2}$$

where θ is the angle of latitude of 1. Notice that the speed of 1 with respect to the coordinate frame (r, θ, ϕ) is

$$R_\oplus \cos\theta \Omega_\oplus = R_\oplus \cos\theta \frac{d\phi}{dt} \quad ,$$

where $\Omega_\oplus$ is the angular velocity of the Earth. Then, we have

---

* Strictly speaking, the Schwarzschild metric should not be valid for a rotating body. However, the influence of the rotation of the Earth on the metric here is quite small, so it is possible to express the gravitational field in the neighbourhood of the Earth by the Schwarzschild metric.

$$\tau_1 = t(1 - \frac{2GM_\oplus}{c^2 R_\oplus} - \frac{R_\oplus^2 \Omega_\oplus^2}{c^2} \cos^2\theta)^{1/2} \quad ,$$

where t is the coordinate time between the initial and final events of synchronisation and comparison respectively.

For clock 2, we have $r = R_\oplus + h$, $dr = 0$, and

$$\tau_2 = t(1 - \frac{2GM_\oplus}{c^2(R_\oplus + h)} - \frac{u^2}{c^2})^{1/2} \quad ,$$

where

$$\vec{u} \cong \vec{v} + \Omega_\oplus (R_\oplus + h) \cos\theta \; \vec{e}_E$$

and $\vec{e}_E$ is the unit vector in the eastward direction. The above equation is only approximate because we have not used the velocity addition formula of relativity.

Now we define

$$\delta = \frac{\tau_2 - \tau_1}{\tau_1} \quad .$$

Substituting τ_1 and τ_2 into δ, and retaining the lowest order of small quantities $GM_\oplus/c^2 R_\oplus$, $h/R_\oplus$, v^2/c^2 and $h\Omega_\oplus/v$, we get

$$\delta = \frac{gh}{c^2} - \frac{v^2}{2c^2} - \frac{\Omega_\oplus R_\oplus \cos\theta}{c^2} \vec{v} \cdot \vec{e}_E \quad , \tag{2.4}$$

where $g = GM_\oplus/R_\oplus^2 = 9.81 \text{ ms}^{-2}$.

The difference between the proper times of similar eastward and westward motions per revolution is

$$\frac{4\pi}{\Omega_\oplus} \left(\frac{\Omega_\oplus R_\oplus \cos\theta}{c} \right)^2 = 4.12 \times 10^{-7} \cos^2\theta \text{ s} \quad ,$$

coming from the last term in (2.4) where

$$|\vec{v}| = 2\pi R_\oplus \cos\theta \text{ per revolution,}$$

the latitudinal distance covered per revolution.

48

By using modern atomic clocks, the above effect is observable. In 1971, Hafele and Keating used two caesium clocks in separate aircraft which flew in opposite directions, eastwards and westwards, with a comparison between the 2 proper times made when the aircraft met again. They did not quite follow formula (2.4) for δ, because the flights were not latitudinal. However, the theory is essentially the same and the results are as shown in Table 2.

Table 2

Direction of circumnavigation	$\tau_2 - \tau_1$ 10^{-9} s	
	Experiment	Theory
Westward	273 ± 7	275 ± 21
Eastward	-59 ± 10	-40 ± 23

These results represent a successful direct test of the dependence of the separation between two events on the world lines pursued by measuring clocks.

2-4   Precession of the Perihelion

It has been said in the first chapter that the precession of the perihelion of Mercury as given by Newtonian mechanics does not agree with observations. Now we will consider a more correct treatment, where the motion of a planet according to general relativity is a time-like geodesic in the Schwarzschild spacetime surrounding the Sun.

Due to the spherical spatial symmetry of the Schwarzschild metric, conservation of angular momentum still holds and the motion of the planet remains in a plane, as is for central motion in Newtonian mechanics. Therefore, we do not lose generality by restricting the motion on the outset to the equatorial plane $\theta = \pi/2$. The world line is then specified by three functions $t(\tau)$, $r(\tau)$ and $\phi(\tau)$, which are determined by the geodesic equation, or more precisely, by extrema values of the following quantity:

$$\tau_{AB} = \int_A^B d\tau \sqrt{- g_{\mu\nu} \frac{dx^\mu}{d\tau} \frac{dx^\nu}{d\tau}} \tag{2.5}$$

$$= \int_A^B d\tau \sqrt{\left(1 - \frac{2GM}{c^2 r}\right)\left(\frac{dt}{d\tau}\right)^2 - \frac{1}{c^2}\left[\left(1 - \frac{2GM}{c^2 r}\right)^{-1} \left(\frac{dr}{d\tau}\right)^2 + r^2\left(\frac{d\phi}{d\tau}\right)^2\right]} \quad ,$$

where A and B are any two events on the planet's world line. (Notice that $M = M_\odot$ in this section).

We could get three equations concerning $t(\tau)$, $r(\tau)$ and $\phi(\tau)$ from this condition, but we need only derive two since we know from the metric expression (2.3) that

$$\left(1 - \frac{2GM}{c^2 r}\right)\left(\frac{dt}{d\tau}\right)^2 - \frac{1}{c^2}\left[\left(1 - \frac{2GM}{c^2 r}\right)^{-1} \left(\frac{dr}{d\tau}\right)^2 + r^2\left(\frac{d\phi}{d\tau}\right)^2\right] = 1 \quad . \tag{2.6}$$

To get equations of motion from (2.5), we imagine a world line W between A and B slightly different from the geodesic world line. Along this world line W, the proper separation is τ_{AB}^W, and the extremum condition would be

$$\delta\tau_{AB} \equiv \tau_{AB}^W - \tau_{AB} = 0 \quad . \tag{2.7}$$

First, we take W to have deviation only in the t coordinate, that is, we assume that the functions $r(\tau)$ and $\phi(\tau)$ describing the planet's path in space are fixed. Thus we have only the variation

$$t(\tau) \to t(\tau) + \delta t(\tau) \quad ,$$

where $\delta t(\tau)$ is any small function of τ that vanishes at A and B. Now if we denote by F the quantity $(- g_{\mu\nu} \frac{dx^\mu}{d\tau} \frac{dx^\nu}{d\tau})^{1/2}$ as a function of t and $\frac{dt}{d\tau}$, then the expression (2.7) becomes

$$\delta\tau_{AB} = \int_A^B d\tau \left[F(t + \delta t, \frac{dt}{d\tau} + \delta\frac{dt}{d\tau}) - F(t, \frac{dt}{d\tau})\right]$$

50

$$= \int_A^B d\tau \left\{ \left[\frac{\delta F(t, \frac{dt}{d\tau} + \delta\frac{dt}{d\tau})\delta t}{\delta t} \right] + \left[\frac{\delta F(t, \frac{dt}{d\tau})}{\delta(\frac{dt}{d\tau})} \right] \delta(\frac{dt}{d\tau}) \right\} \quad .$$

After integrating by parts, we obtain

$$\delta\tau_{AB} = \int_A^B \left[\frac{\partial F}{\partial t} - \frac{d}{d\tau} \frac{\partial F}{\partial(\frac{dt}{d\tau})} \right] d\tau \, \delta\tau = 0 \quad .$$

Since $\delta t(\tau)$ is an arbitrarily small function, the integral can vanish only if

$$\frac{d}{d\tau} \frac{\partial F}{\partial(\frac{dt}{d\tau})} - \frac{\partial F}{\partial t} = 0 \quad .$$

Using the explicit form for F, we get

$$(1 - \frac{2GM}{c^2 r}) \frac{dt}{d\tau} = \text{const} \equiv \frac{E}{mc^2} \quad , \tag{2.8}$$

where the right-hand side can be deduced from (2.6). By an analogous procedure, we can find the equation obtained by varying $\phi(\tau)$, with the result:

$$r^2 \frac{d\phi}{d\tau} = \text{const} \equiv L_\phi/m \quad . \tag{2.9}$$

The constants E in (2.8) and L_ϕ in (2.9) are respectively the energy and ϕ components of angular momentum of an orbiting body of mass m.

Now, eqs. (2.6), (2.8) and (2.9) are the equations of motion of the geodesic, from which we can get

$$(\frac{dr}{d\phi})^2 = \frac{r^4}{L_\phi^2} \left[(\frac{E}{c})^2 - (1 - \frac{2GM}{c^2 r})(\frac{L_\phi^2}{r^2} + m^2 c^2) \right] \quad .$$

Let $u = 1/r$, and the above equation can be rewritten as

$$(\frac{du}{d\phi})^2 + u^2 = \frac{(E/c)^2 - m^2 c^2}{L_\phi^2} + \frac{m^2 c^2}{L_\phi^2} (\frac{2GM}{c^2}) u + (\frac{2GM}{c^2}) u^3 \quad .$$

By differentiating with respect to ϕ, noting that $u = u(\phi)$, we get

51

$$\frac{d^2u}{d\phi^2} + u = \frac{GM}{a^2} + \frac{3}{2} \left(\frac{2GM}{c^2}\right) u^2 \quad , \tag{2.10}$$

where $a = L_\phi/m$. In a weak field, $\left(\frac{2GM}{c^2}\right) u \ll 1$, and the second term on the right-hand of (2.10) can then be neglected in the zero-order. In such a case the solution is

$$u_0 - \frac{GM}{a^2} = A \cos(\phi + \phi_0) \quad , \tag{2.11}$$

where A and ϕ_0 denote the constants of integration. The orbits of (2.11) are conic sections and are specified in terms of eccentricity $e = Aa^2/GM$, and perihelion distance $r_{min} = a^2/GM(1 + e)$. If $e < 1$, the orbits are bound and elliptical in shape (Fig. 2.3). In the case for which the minor axis is parallel to $\phi = 0$ (i.e. $\phi_0 = 0$), the ellipse can be written as

$$u_0 = \frac{1}{r} = \frac{GM}{a^2} (1 + e \cos\phi) \quad . \tag{2.12}$$

We shall calculate the correction to the elliptical orbits caused by the relativistic term $\frac{3}{2} \left(\frac{2GM}{c^2}\right) u^2$ in (2.10). The value of this term is only about 10^{-7} for Mercury and far less for other planets, so that it is only necessary to calculate the lowest order corrections, called

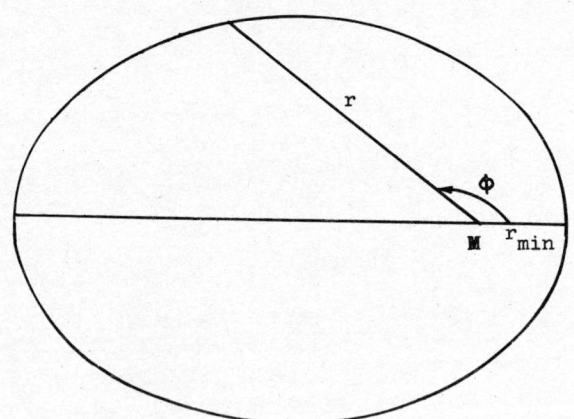

Fig. 2.3 Orbit of a planet

the post-Newtonian corrections. Substituting (2.12) into the second
term on the right-hand side of (2.10), we get

$$\frac{d^2u}{d\phi^2} + u = \frac{GM}{a^2} + \varepsilon \, \frac{3GM}{a^2} \, [2e \cos\phi + (1 + e^2 \cos^2\phi)] \quad ,$$

where $\varepsilon = (GM/ca)^2 \ll 1$. Let $u = u_0 + u_1$. Then the equation for the
first-order correction function u_1 is

$$\frac{d^2u_1}{d\phi^2} + u_1 = \varepsilon \, \frac{3GM}{a^2} \left[2e \cos\phi + (1 + e^2 \cos^2\phi) \right] \quad . \qquad (2.13)$$

This is an equation for forced oscillations. In (2.13), the only
important term on the right-hand side is the first one, which is res-
onant, while the second non-resonant term will only cause a slight
periodic variation in the position of the perihelion. Thus, after
neglecting the non-resonant term, equation (2.12) becomes

$$\frac{d^2u_1}{d\phi^2} + u_1 = \varepsilon \, \frac{6GM}{a^2} e \cos\phi \quad .$$

A solution can be obtained as

$$u_1 = \varepsilon \, \frac{3GMe}{a^2} \phi \sin\phi \quad .$$

It is obvious that the presence of a multiplicative factor ϕ in the
solution causes a cumulative effect which can be observed clearly after
a sufficiently large number of revolutions.

Using the above solution, by considering the relativistic correction up to the first order, the orbit is

$$u = u_0 + u_1 = \frac{GM}{a^2} [1 + e(\cos\phi + 3\varepsilon\phi\sin\phi)]$$

or

$$r \simeq \frac{a^2}{GM} \Big/ \left\{ 1 + e \cos [\phi(1 - 3\varepsilon)] \right\} \quad ,$$

as ε is small.

We know that perihelia occur when the cosine is unity, and are
therefore determined by the following equation:

$$\phi(1 - 3\varepsilon) = 2\pi n \quad ,$$

where n is any integer. This can be approximated as

$$\phi = 2\pi n + 6\pi n \epsilon \quad .$$

Therefore, the azimuth angle ϕ increases with increasing n, that is, the major axis of the ellipse has a precession. The angular precession $\Delta\phi_1$ per revolution is

$$\Delta\phi_1 = \frac{6\pi GM}{c^2 r_{min}(1 + e)} \quad , \tag{2.14}$$

and the centennial precession $\Delta\phi$ is

$$\Delta\phi = \frac{6\pi GMN}{c^2 r_{min}(1 + e)} \quad ,$$

where N is the number of revolutions per century.

Only for the planets Mercury, Venus and the Earth, and the asteroid Icarus, is r_{min} small enough and M large enough for $\Delta\phi$ to be measured. The results are as shown in Table 3. The large uncertainty in the measured precession of Venus arises from the near-circularity of the orbit (e is only 0.0068), which makes it difficult to locate the precession. These results support the verification of general relativity.

Table 3

| Planet | $\Delta\phi^{100}$ (seconds of arc) | |
	Observation	Theory
Mercury	43.11 ± 0.45	43.03
Venus	8.4 ± 4.8	8.6
Earth	5.0 ± 1.2	3.8
Icarus	9.8 ± 0.8	10.3

To confirm that all of the remaining precession of a star arises from general relativity, it is of course necessary to be able to rule out other possibilities which may also cause some precession, the most important of which is the non-spherical symmetry of the Sun. If the Sun is slightly oblate, its gravitational potential would be

$$V = \frac{GM}{r} \left[1 - J_2 \frac{R_\odot^2}{r^2} \left(\frac{3 \cos^2\theta - 1}{2} \right) \right] \quad ,$$

where J_2 is the oblateness of the Sun. In this field, there is already a certain rotation of the perihelion, the value of which per revolution of the star would be

$$\Delta\phi = \frac{6\pi GM}{r_{min} (1 + e)} J_2 \frac{R_\odot^2/MG}{2r_{min} (1 + e)} \quad .$$

Thus our ignorance of J_2 is the outstanding serious problem that prevents us from isolating the relativistic contribution to the precession. Inference of J_2 from measurements of the visual oblateness of the Sun is difficult; this method has been tried, but the results are in dispute. Dicke and Goldenberg have claimed that this oblateness is as large as $J_2 = 5 \times 10^{-5}$, which should account for about 20% of the remaining precession. However, recent observations indicate that the oblateness of the Sun is far less than the above value with only $J_2 = (1.84 \pm 1.25) \times 10^{-6}$. Inference of J_2 by comparing results for Mercury and Mars is also difficult. The effect for Mars is very small, and the influences of the asteroid belt on the orbit of Mars make the interpretation of a measured precession difficult.

The best approach for measuring J_2 would be to track a spacecraft that passes close to the Sun. In one possible version of such a method, the spacecraft would be sent from the Earth to pass by Jupiter to obtain a "gravity assist". Due to the Jupiter encounter, the spacecraft would be made to travel perpendicular to the ecliptic. After several years of flight, the spacecraft would pass by the Sun in less than a day and J_2 would be estimated from that brief encounter.

2-5   Deflection of Light

The principle of equivalence implies that light is deflected in a gravitational field. For instance, imagine a laboratory falling freely in the gravitational field near the Earth's surface, and suppose a light ray is emitted from the left end propagates horizontally towards the right (Fig. 2.4). According to the principle of equivalence, the laboratory is an inertial frame in which gravity is eliminated, and the light ray is observed to travel along a straight line according to the demands of special relativity. However, for an observer on the ground, the laboratory is accelerated downwards, and the light ray which propagates horizontally to the right for an observer in the laboratory, would in this frame be accelerated downwards in a curved path. Due to the weakness of the gravitational field near the Earth's surface, deflection of the light is very small (a light ray propagating horizontally has a deflection of only about 1Å, due to a curved trajectory, for every kilometre travelled) and difficult to observe. Nevertheless, the deflection of light in the Sun's gravitational field has a magnitude that can be observed. For a calculation of the total deflection caused by the Sun's gravitational field, it is necessary that in obtaining a correct result we must consider many different local inertial frames and their connections. It is not sufficient to employ just the

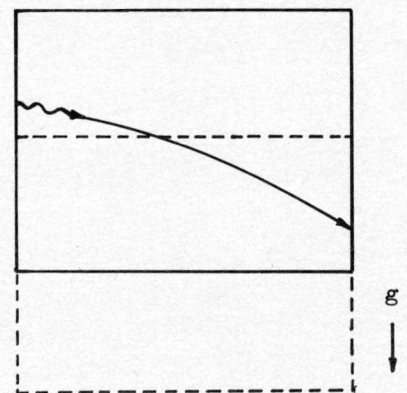

Fig. 2.4   The light ray is deflected for
an observer on the ground

single laboratory falling freely as mentioned above.

Before giving a treatment based on general relativity, let us analyse the problem using Newtonian mechanics. As early as in 1801, Soldner calculated the deflection of light in gravitational fields using Newtonian mechanics.

The formula (2.12) corresponds to unbound hyperbolic orbits if the eccentricity e exceeds unity. The asymptotes, where $r \to \infty$, correspond to angles (Fig. 2.5)

$$\phi_{\pm} = \pm \left(\frac{\pi}{2} + \frac{1}{2} \delta \right)$$

where δ is the total Newtonian deflection of the ray, given by

$$\cos \phi_{\pm} = -\frac{1}{e} \ ,$$

i.e. $\sin \frac{1}{2} \delta = \frac{1}{e}$.

Since the speed of light is c, by considering (2.9) and $a = L_{\phi}/m$, we can see that

$$a \simeq r_{min} c \ .$$

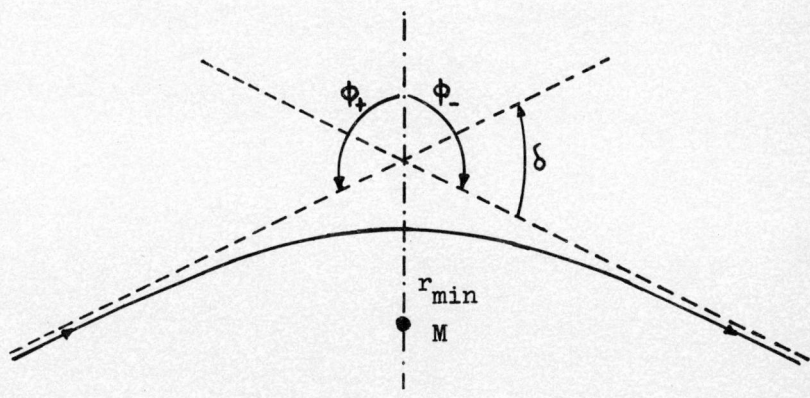

Fig. 2.5 The deflection of light in the gravitational field of the Sun

This enables us to find the eccentricity via

$$r_{min} = \frac{a^2}{GM(1 + e)} \quad ,$$

which is

$$e = \frac{a^2}{GMr_{min}} - 1 \simeq \frac{c^2 r_{min}}{GM} - 1 \simeq \frac{c^2 r_{min}}{GM} \quad .$$

Since $c^2 r_{min}/GM \gg 1$, e is very large and δ is very small, so that we have approximatively,

$$\sin \left(\tfrac{1}{2}\delta\right) \simeq \tfrac{1}{2}\delta = \frac{1}{e} \quad ,$$

that is

$$\delta = \frac{2GM}{c^2 r_{min}} \quad .$$

For light grazing the surface of the Sun, $r_{min} = R_\odot$ and $M = M_\odot$, giving

$$\delta = 0''.875 \quad .$$

Now let us calculate the deflection δ according to general relativity. The results obtained in the last section can be used for discussing light propagation in gravitational fields, but it should be noticed that the rest mass of light is zero. Taking $m = 0$, eq. (2.10) becomes

$$\frac{d^2 u}{d\phi^2} + u = \frac{3GM}{c^2} u^2 \quad . \tag{2.15}$$

If $M = 0$, the path of the light would be a straight line with the orbit equation,

$$u_0 = \frac{1}{r_{min}} \cos \left(\phi + \phi_0\right) \quad ,$$

where r_{min} and ϕ_0 are integral constants. By making $\phi_0 = 0$, up to the first order correction, eq. (2.15) gives

$$\frac{d^2u}{d\phi^2} + u = \frac{3GM}{c^2 r_{min}^2} \cos^2\phi \quad ,$$

which has a solution:

$$u = \frac{1}{r_{min}} \cos\phi + \frac{GM}{c^2 r_{min}^2} (1 + \sin^2\phi) \quad .$$

The asymptote is determined by taking $r \to \infty$, namely,

$$0 = - \frac{1}{r_{min}} \sin \frac{\delta}{2} + \frac{GM}{c^2 r_{min}^2} (1 + \cos^2 \frac{\delta}{2}) \quad .$$

Since $\delta \sim 0 \, (GM/c^2 r_{min}) \ll 1$, we get the deflection δ as

$$\delta \simeq \frac{4GM}{c^2 r_{min}} \quad . \tag{2.16}$$

This is twice the Newtonian value. For light grazing the Sun,

$$\delta = 1".75 \quad .$$

It is only possible to measure the deflection of light from a star during a total eclipse of the Sun. Measuring the relative positions of the stars around the Sun during an eclipse and repeating the measurements for the same celestial region six months later (i.e. in the absence of the Sun's gravitational field in the region), a comparison between the two results would give the required deflection. In this way $\Delta\phi$ has been measured for about 400 stars since 1919. The experimental results all lie within the limits $1".57-2".37$, the mean value being $1".89$. These results disagree with the prediction of Newtonian theory but agree quite well with general relativity.

There are many problems that are difficult to overcome in the observation of deflections. First of all, the effect of the solar corona limits us to measurements of the star with $r_{min} > 2R_\odot$. Secondly, total eclipses of the Sun are not usually observable at locations where large telescopes are available. The size of the diffraction disc of a telescope of 10 cm in diameter is about 5×10^{-6} arc, which restricts the accuracy of the measurement. Moreover, exposures and developing which are made at different times also bring in systematic errors.

Radiosources have been employed for detecting the deflection of light in the last few years. The results of direction measurements using very long baseline interferometry are better than those using the optical method, since the precision of the former can be very high. For example, QSO 3C279 is occulted annually by the Sun. By measuring the angle between 3C279 and 3C273, before and after an occultation, the required results on the deflection are obtained. Some of these results are listed in Table 4.

Table 4

Name of observatory	Frequency (MHz)	Length of baseline (km)	$\Delta\phi$
OWENSVALLEY	9602	1	1".7 ± 0".20
GOLDSTONE	2388	21.566	1".82 ± 0".24 0".17
GOLDSTONE HAYSTACK	7840	3899.22	1".80 ± 0".2
NRAO	2695 8085	2.7	1".57 ± 0".08
NRAO	2697 4993.8	1.41	1".87 ± 0".3

In addition, radiosources 0119+11, 0116+08 and 0111+02 are collinear so that when the ecliptic of the Sun crosses 0116+08, 0119+11 and 0111+02 are each on one side of the ecliptic, making angles of 4^0 and 6^0 with the ecliptic respectively. The Sun passes through the celestial region near 0116+08 in the first ten days of the month every April. Using two frequencies, 2695 and 8085 MHz, eliminates the effects of the corona. Fomaleont and Sramek have measured the change in the relative positions of the three radiosources by using 35 km baseline interferometry at NARO when the Sun passed 0116+08. Their result is $\Delta\phi = 1".761 \pm 0".010$.

2-6 Experiments on Radar Echoes

Measurements on the deflection of light have been used to test the predictions of general relativity concerning the shape of the orbits of light rays. The time coordinate along the orbit is eliminated to obtain an equation for r in terms of ϕ. However, experiments using so-called radar echoes enable tests on the predictions involving the time coordinate to be made. The method in such experiments is as follows: a radar signal is emitted towards a planet and the echo reflected back by the planet is received on the Earth. If no effect of the gravitational fields of the Sun were to exist, the radar signal would travel along a straight line with the speed of light c. When the path of the signal is chosen to pass near the Sun, the echo should be delayed relative to the former case due to the effect of the Sun's gravitational fields. This means that the signal slows down as it passes by the Sun.

Now let us use directly the Schwarzschild metric to calculate the coordinate speed of light travelling in the (r, ϕ) plane defined by $\theta = \pi/2$. For light, $d\tau = 0$, so that eq. (2.3) now becomes

$$1 - \frac{2GM}{c^2 r} - \frac{1}{c^2}\left[(1 - \frac{2GM}{c^2 r})^{-1}(\frac{dr}{dt})^2 + r^2(\frac{d\phi}{dt})^2\right] = 0 \quad . \tag{2.17}$$

From (2.8) and (2.9),

$$r^2(\frac{d\phi}{dt}) = \frac{L_\phi}{m}(\frac{mc^2}{E})(1 - \frac{2GM}{c^2 r}) \quad ,$$

which in (2.17) gives

$$\frac{1}{c^2}(\frac{dr}{dt})^2 = \left(1 - \frac{2GM}{c^2 r}\right)^2 - \ell^2\left(\frac{2GM}{c^2 r}\right)^2\left(1 - \frac{2GM}{c^2 r}\right)^3 \quad ,$$

where $\ell = L_\phi c^3/2EGM$. The factor ℓ can be replaced in terms of a small parameter r_o, which is defined by $dr/dt = 0$ in the above equation, giving

$$\ell = \frac{c^2 r_o}{2GM}\left(1 - \frac{2GM}{c^2 r_o}\right)^{-\frac{1}{2}} \quad .$$

61

Thus we have

$$\frac{1}{c^2}\left(\frac{dr}{dt}\right)^2 = \left(1 - \frac{2GM}{c^2 r}\right)^2\left[1 - \left(\frac{r_0}{r}\right)^2 \frac{1 - (2GM/c^2 r)}{1 - (2GM/c^2 r_0)}\right] \equiv F(r) \quad .$$

The coordinate time t_{AB} between the emission of the signal and the reception of the echo (Fig. 2.6) is given by

$$t_{AB} = \frac{2}{c}\int_A^B \frac{dr}{\sqrt{F(r)}} \quad .$$

Since $2GM/c^2 r \ll 1$, $\sqrt{F(r)}$ can then be expressed as

$$\sqrt{F(r)} \simeq \left(1 - \frac{2GM}{c^2 r}\right)\left\{1 - \left(\frac{r_0}{r}\right)^2 \left(1 - \frac{2GM}{c^2 r} + \frac{2GM}{c^2 r_0}\right)\right\}^{1/2}$$

$$= \left[1 - \left(\frac{r_0}{r}\right)^2\right]^{1/2}\left(1 - \frac{2GM}{c^2 r}\right)\left\{1 - \frac{2GMr_0}{c^2(1 + \frac{r_0}{r})r^2} - \frac{r_0}{r} \cdots\right\}^{1/2}$$

$$\simeq \left[1 - \left(\frac{r_0}{r}\right)^2\right]^{1/2}\left(1 - \frac{2GM}{c^2 r} - \frac{GMr_0}{c^2(r + r_0)r}\right) \quad .$$

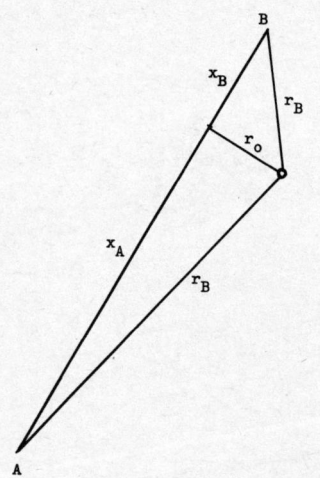

Fig. 2.6 Experiment on radar echo

62

Thus

$$\Delta t \equiv t_{AB} - \frac{2}{c} \int_A^B \frac{dr}{\sqrt{1 - (r_0/r)^2}}$$

$$\simeq \frac{4GM}{c^3} \left\{ \int_A^B \frac{dr}{(r^2 - r_0^2)^{1/2}} + \frac{1}{2} \int_A^B \frac{r_0}{(r^2 - r_0^2)^{1/2}(r + r_0)} \, dr \right\} \, .$$

$$(2.18)$$

The term $\dfrac{2}{c} \displaystyle\int_A^B \dfrac{dr}{\sqrt{1 - (r_0/r)^2}}$ is the time needed for light to travel

A-B-A in the absence of the gravitational field of M. Therefore, Δt is the relativistic correction and the right-hand side of (2.18) is the correction of the post-Newtonian order. The integration can be done by using the substitution $r = r_0 \sec\phi$, with the result:

$$\Delta t \cong \frac{4GM}{c^3} \left\{ \int \frac{d\phi}{\cos\phi} + \frac{1}{2} \int \frac{d\phi}{1 + \cos\phi} \right\}$$

$$= \frac{2GM}{c^3} \left[2\ln \frac{(r_A + x_A)(r_B + x_B)}{r_0^2} - \left(\frac{r_0}{x_A} + \frac{r_0}{x_B} \right) + \left(\frac{r_A}{x_A} + \frac{r_B}{x_B} \right) \right] \, ,$$

where x_A and x_B are shown in Fig. 2.6. In practical cases, $M = M_\odot$, r_0 is the impact parameter of the signal which is of the order of $R_\odot$, and $r_A = r_\oplus$, $r_B = r_p$ are respectively the coordinate distances of the Earth and the planet from the Sun. Therefore, we have $r_\oplus$, $r_p \gg r_0$, and the correction terms can be simplified further. The final result is

$$\Delta t \simeq \frac{4GM}{c^3} \left(\ln \frac{4r_\oplus r_p}{r_0^2} + 1 \right) \, . \tag{2.19}$$

For $r_0 = R_\odot$, Δt for the Earth-Mercury echo is about 2.4×10^{-4} s. It should be noted that t_{AB} is the coordinate time between two events on the Earth, that is, at essentially fixed values of (r, θ, ϕ). The corresponding proper time, which is what our clocks would measure, is smaller by a factor $(1 - 2GM/c^2 r_\oplus)^{1/2}$, differing from unity by only 10^{-8}, and can thus be neglected with respect to Δt.

63

It can be seen from (2.19) that when the light path passes close to the Sun, the argument of the logarithmic term gives rise to the characteristically sharp spike which is very important because it makes this effect relatively easy to distinguish from the various perturbations of the orbits of the Earth and the target planet.

Several experiments have been performed since 1964, principally by Shapiro et al. In every case, the predictions of general relativity were confirmed. One example of the results, shown in Fig. 2.7, was made for the superior conjunction of Venus in 1970.

From radar data obtained throughout 1972, it was possible to get the result with an uncertainty of 4%. The main contribution to the uncertainty was the limited signal-to-noise ratio available. The solar corona had a negligible effect since the radio frequency used in the radar measurements was nearly 8 GHz.

One point deserves emphasis: the results of radar echoes are completely incompatible with the predictions of Newtonian mechanics, since, according to Newtonian mechanics, light should travel faster near the Sun, Δt should be negative in all practical cases.

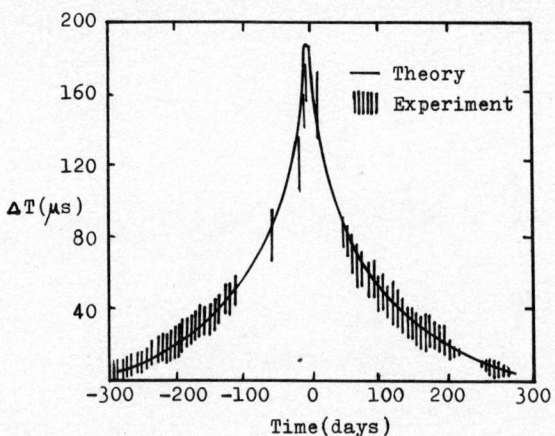

Fig. 2.7 Radar time delay of signals reflected from Venus and passing close to the Sun: Comparison of theory with experiment.

2-7   Precession of the Axis of Rotation

In Newtonian theory, gravity does not influence the rotation of a body, and the direction of the axis of rotation of a body moving in the gravitational field of the Earth is therefore independent of the motion of the centre of mass of the body. The result from general relativity is different due to the intrinsic rotation to orbit motion coupling, which can be considered as an SL coupling of general relativity.

Let us discuss the coupling related to rotation in the case of the two-body problem. The two-body problem can be solved rigorously as the well-known Kepler problem in Newtonian mechanics. In general relativity, however, the two-body problem, which is different from the problem of a planet's orbit discussed in section 2.4, is too difficult to be solved precisely. In section 2.4, we only considered the planet as a test body in the gravitational field of the Sun and neglected the effect of the planet on the gravitational field itself. Thus, the problem becomes one of calculating the geodesic in the Schwarzschild metric. However, when the masses of the two bodies are comparable, this approximation would not be available, which is why this problem is difficult to solve. We will only give the physical results below without its deductive procedure.

For a binary system that may consist of stars 1 and 2, we use m_1 and m_2, $\vec{r}_1$ and $\vec{r}_2$, $\vec{v}_1$ and $\vec{v}_2$, $\vec{p}_1$ and $\vec{p}_2$, $\vec{s}^{(1)}$ and $\vec{s}^{(2)}$, $\vec{n}^{(1)}$ and $\vec{n}^{(2)}$ to express mass, position, velocity, momentum, angular momentum of self-rotation (intrinsic rotation or spin) and the unit vectors of the directions of the intrinsic angular momentum of the stars 1 and 2 respectively. We will use $\vec{r} \equiv \vec{r}_1 - \vec{r}_2$, reduced mass $\mu \equiv m_1 m_2/(m_1 + m_2)$ and total mass $M \equiv m_1 + m_2$.

The following relativistic effects do not appear in Newtonian binary systems:

1. <u>Precession of Periastron</u>

This effect is similar to the precession of perihelia discussed in section 2.4. Since two-body problems in Newtonian theory are equivalent to the motion of a test body in a central field, the bounded

65

orbit $\vec{r}(t)$ is an ellipse and the direction of the periastron (nearest point of the two stars in binary) does not change. One of the relativistic effects is a precession of the axis of the elliptical orbit.

The Runge-Lenz vector $\vec{A}$ is generally employed for treating this problem. The definition of $\vec{A}$ is as follows:

$$\frac{\vec{A}}{\mu} = \vec{v} \times (\vec{r} \times \vec{v}) - \frac{GM\vec{r}}{R} \quad ,$$

where $\vec{v} = d\vec{r}/dt$. It can be proved that, for the two-body problem in Newtonian mechanics, $\vec{A}$ is an integral of motion. The direction is such that $\vec{A}$ points exactly at the periastron of the elliptical orbit $\vec{r}(t)$. The fact that $\vec{A}$ is a constant vector indicates that the direction of the periastron is unchanging.

For the two-body problem in general relativity, $\vec{A}$ must satisfy the following equation:

$$\frac{d\vec{A}}{dt} = \vec{\Omega} \times \vec{A} \tag{2.20}$$

which states that $\vec{A}$ precesses with a resultant angular velocity $\vec{\Omega}$ given by

$$\vec{\Omega} = \vec{\Omega}^{(E)} + \vec{\Omega}^{(1)} + \vec{\Omega}^{(2)} + \vec{\Omega}^{(1,2)} \quad , \tag{2.21}$$

where

$$\vec{\Omega}^{(E)} = A^{(E)}\vec{n}$$

$$\vec{\Omega}^{(1)} = A^{(1)}[\vec{n}^{(1)} - 3(\vec{n}\cdot\vec{n}^{(1)})\vec{n}]$$

$$\vec{\Omega}^{(2)} = A^{(2)}[\vec{n}^{(2)} - 3(\vec{n}\cdot\vec{n}^{(2)})\vec{n}]$$

$$\vec{\Omega}^{(1,2)} = A^{(1,2)}\left\{(\vec{n}\cdot\vec{n}^{(1)})\vec{n}^{(2)} + (\vec{n}\cdot\vec{n}^{(2)})\vec{n}^{(1)}\right.$$
$$\left. + [\vec{n}^{(1)}\cdot\vec{n}^{(2)} - 5(\vec{n}\cdot\vec{n}^{(1)})(\vec{n}\cdot\vec{n}^{(2)})]\vec{n}\right\}$$

$$A^{(E)} = \frac{3GM\bar{\omega}}{c^2 a(1 - e^2)}$$

$$A^{(1)} = \frac{Gs^{(1)}(4 + 3m_2/m_1)}{2c^2 a^3 (1 - e^2)^{3/2}}$$

$$A^{(2)} = \frac{Gs^{(2)}(4 + 3m_1/m_2)}{2c^2 a^3 (1 - e^2)^{3/2}}$$

$$A^{(1,2)} = -\frac{3Gs^{(1)}s^{(2)}/\mu\bar{\omega}}{2c^2 a^5 (1 - e^2)^2} \quad .$$

Here, $\bar{\omega}$ is the mean orbital angular-velocity, a is half of the major axis, and $\vec{n}$ is the unit vector in the direction of the orbital angular momentum. If we let $s^{(1)}$, $s^{(2)}$ and m_1 in the above formula be zero, then only $\vec{\Omega}^{(E)}$ is non-vanishing and equals (2.14).

2. Precession of Rotation Axis

The rotational direction of body 1 obeys the following equation:

$$\frac{d\vec{n}^{(1)}}{dt} = \vec{\Omega}^{(1)} \times \vec{n}^{(1)} \quad . \tag{2.22}$$

This is also a precession equation, and indicates that the rotational direction $\vec{n}^{(1)}$ of body 1 precesses with an angular velocity $\vec{\Omega}^{(1)}$ given by

$$\vec{\Omega}^{(1)} = \vec{\Omega}_D^{(1)} + \vec{\Omega}_L^{(1)} \quad , \tag{2.23}$$

with

$$\vec{\Omega}_D^{(1)} = A_D^{(1)}\vec{n}$$

$$\vec{\Omega}_L^{(1)} = A_L^{(1)}[\vec{n}^{(2)} - 3(\vec{n}\cdot\vec{n}^{(2)})\vec{n}]$$

$$A_D^{(1)} = \frac{3G\bar{\omega}(m_2 + \mu/3)}{2c^2 a(1 - e^2)}$$

$$A_L^{(1)} = \frac{Gs^{(2)}}{2c^2 a^3 (1 - e^2)^{3/2}} \quad .$$

When we take $c \to \infty$, then $\vec{\Omega}^{(1)} = 0$, which means that the body's rotation in Newtonian mechanics would not be influenced by gravity.

67

Generally,the orbital angular momentum of a system of binary stars is always much larger than that of spin rotation, that is, $\mu\bar{\omega}a^2(1 - e^2)^{1/2} \gg s^{(1)}$, $s^{(2)}$. Therefore, the precession of the periastron is principally from $\vec{\Omega}^{(E)}$, and the precession of the rotation axis, from $\vec{\Omega}_D^{(1)}$ where $\vec{\Omega}^{(E)}$ and $\vec{\Omega}_D^{(1)}$ have same direction and the ratio of their magnitudes is

$$\frac{|\vec{\Omega}^{(1)}|}{|\vec{\Omega}^{(E)}|} = \frac{(m_2 + \mu/3)}{2M} = \frac{m_2(4m_1 + 3m_2)}{6(m_1 + m_2)^2} \quad .$$

Now let us discuss the observational test of SL coupling in general relativity. In 1974, Hulse and Taylor found a new radio pulsar PSR 1913+16 (for details concerning the pulsar, see the next chapter), which differs from other known radio pulsars. This pulsar is a member of a binary system with a very short period of revolution (less than 8 hours), which means that $\bar{\omega}$ is large and the distance a between the two stars is very small. Hence, $\Omega^{(E)}$ of the binary system must be large. The precession rate of its periastron has indeed been found to be

$$\Omega^{(E)} = 4°.226 \pm 0°.002/\text{year} \quad ,$$

which is greater than the precession of Mercury by a factor of several ten thousands.

It was found that the masses of the two stars in the binary systems are about the same, $m_1 \approx m_2 \approx 1.4 \text{ M}_\odot$. Thus, from (2.23), the precession rate due to SL coupling of general relativity is:

$$|\vec{\Omega}^{(1)}| \approx \frac{7}{24} |\vec{\Omega}^{(E)}| \approx 1°/\text{year} \quad .$$

Obvious changes have taken place in the pulse profile of PSR 9113+16 since its discovery. The curves of the pulse profile for several different periods are shown in Fig. 2.8.

We know that the radiation of a pulsar appears to be conic, the section of which is shown in Fig. 2.9. Radiation in the surface and central regions is stronger. When the conical beam of radiation sweeps

68

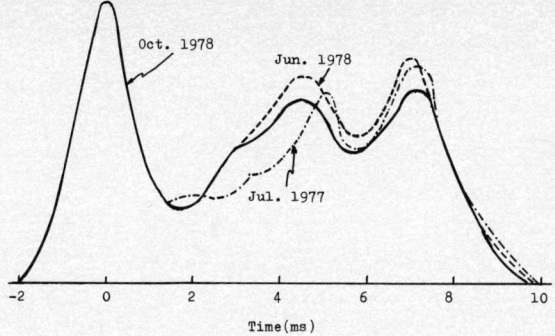

Fig. 2.8   Profile of the pulse of the radio pulsar
PSR 1913+16 at several different times

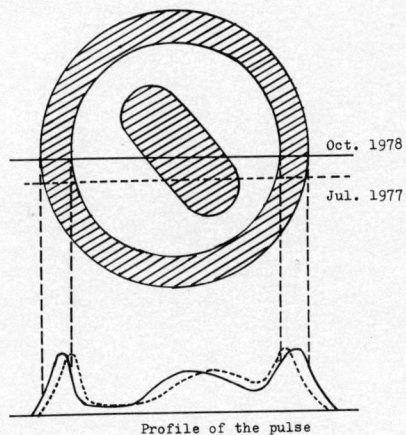

Profile of the pulse

Fig. 2.9   A possible mechanism of changing profile of pulse.
The shadows indicate the section of the radiation
conic of the pulsar, the horizontal lines are the
orbits swept by the line of sight while the pulsar
rotates.

the Earth as the pulsar rotates, a pulse is observed from which the profile is obtained. If the rotation axis precesses, then the way it sweeps the Earth would be different at different times. The two horizontal lines are two possible ways of sweeping the Earth as recorded in July 1977 and October 1978 respectively. Thus, as changes of the pulse profile have actually been observed, we may conclude that such changes are probably consequences of the precession of the rotation axis. The observed result agrees qualitatively with the theoretical prediction of a precession rate of 1°/year.

REFERENCES

Experiment of Gravitational Redshifts:

R.V. Pound and G.A. Rebka, Phys.Rev.Lett. 4, 337 (1960).

R.V. Pound and J.L. Snider, Phys.Rev.Lett. 13, 539 (1964).

R.F.C. Vessot et al, Phys.Rev.Lett. 45, 2081 (1980).

Precession of Planetary.Perihelia:

R.L. Duncombe, Astron.J. 61, 174 (1956).

I.I. Shapiro, C.C. Counselman, R.W. King, Phys.Rev.Lett. 36, 555 (1976).

Gravitational Deflection of Light:

H. Von Kluber, Vistas in Astronomy 3, 47 (1960).

E.B. Fomaleont, R.A. Sramek, Phys.Rev.Lett. 36, 1475 (1976).

Delay of Radar Echoes:

I.I. Shapiro, Phys.Rev.Lett. 26, 1132 (1972).

Spin-Orbit Coupling:

J.H. Taylor, L.A. Fowler, P.H. McCulloch, Nature 277, 437 (1979).

Chapter 3

COMPACT STARS

3-1 Historical Records

Aristotelian theory occupied a dominant position in Europe for a very long historical period. The essential viewpoint of Aristotle in astronomy was that of an unchanging universe, consisting of a series of transparent sphere-like crystals with the Earth at its centre; the spheres revolve around the Earth with different angular velocities, and planets and stars are inlaid on their respective spheres.

Governed by this idea, one was unlikely to think that there would be any anomalous phenomena in the sky, let alone look out for any celestial changes, and if by chance anomalies were seen, one would have had to turn a blind eye on what was observed. This is exactly the reason why there is a shortage of records concerning changes of astronomical phenomena in the historical literature of Europe. A historian of science, Sarton, noted that Europeans and Arabs during the Middle Ages could not understand changing celestial phenomena, not because they had trouble in observing them, but because they believed blindly in the prejudicial view of the perfection of celestial bodies, which were absolutely inert in their minds.

Thus, there were no relevant European records until the Renaissance. The supernova observed by Tycho in 1572 played a very

71

important role in developing scientific ideas (at that time new stars were not yet part of the cosmology) in Europe. Tycho wrote:

During my walk contemplating the sky here and there, since the clearer sky seemed to be just what could be wished for in order to continue observations after dinner, behold, directly overhead, a certain strange star was suddenly seen, flashing its light with a radiant gleam and it struck my eyes. Amazed, and as if astonished and stupefied, I stood still, gazing for a certain length of time with my eyes fixed intently upon it and noticing that the same star was placed close to the stars which antiquity attributed to Cassiopeia. When I had satisfied myself that no star of that kind had ever shone forth before, I was led into such perplexity by the unbelievability of the object that I began to doubt the faith of my own eyes, and so, turning to the servants who were accompanying me, I asked them whether they too could see a certain extremely bright star when I pointed out the place directly overhead. They replied immediately with one voice that they saw it completely and that it was extremely bright. But despite their affirmation, still being doubtful on account of the novelty of the object, I inquired some country people who by chance were travelling past in carriages whether they could see a certain star in the heights. Indeed, these people shouted out that they saw that huge star, which had never been noticed so high up. And at length, having confirmed that my vision was not deceiving me, but in fact that an unusual star existed there, beyond all types, and marvelling that the sky had brought forth a certain new phenomenon to be compared with the other stars.

In comparison, Chinese tradition is quite different. Many of the ancient Chinese astronomers did not deny the viewpoint of the changeability of celestial phenomena and maintained that celestial bodies could often change. When he arrived in China, the Italian missionary Matteo Ricci was surprised by these ideas. He wrote in a letter:

They (the Chinese) say that there is only one sky and not ten skies; that it is empty and not solid. The stars are supposed to move in the void, instead of being attached to the firmament.

72

A wealth of historical information concerning celestial objects in ancient Chinese texts and records are of very great value to us. As early as three or four thousand years ago, there were already records of this type in inscriptions on bones or tortoise shells of the Shang Dynasty. Figure 3.1 shows an inscribed specimen that states that a new star appeared near Antares on the evening of the seventh day.

The sudden appearance of a bright star that was never observed before must have been spectacular, and it was natural to call it a "new star". The name "Nova" has been used in Europe for this kind of star since modern times. In China, such stars were said to be members of so-called "guest" stars, because they usually appeared with an intensity visible to the naked eye for only a short while, to disappear like a visitor pressed for time.

Their lifetimes, between appearance and disappearance, differ from only several days for some to several months for others.

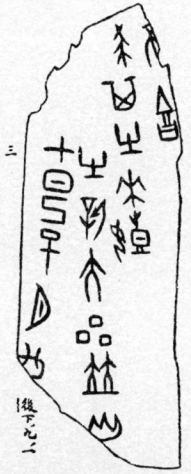

Fig. 3.1   The oldest existing written record
of a nova. These inscriptions are
dated to be of about the 13th century
B.C. The characters state: "On the
evening of the seventh day, a new star
appeared near Antares."

73

It was very rare to observe a nova with a life time exceeding
half a year. After investigating the historical records of two thousand
years, we have only found eight novae with lifetimes greater than six
months. We call them supernovae, which are worthy of special attention.
The general data on the eight supernovae are listed in Table 5.

Table 5

Year (A.D.)	Constellation	Magnitude	Lifetime	Original Record(s)
185	Centaurus	-8	20 months	China
393	Scorpio	-1	8 months	China
1006	Lupus	-8 —— -10	several years	China Japan Europe Arabia
1054	Taurus	-5	22 months	China Japan
1181	Cassiopeia	0	6 months	China Japan
1408	Cygnus	-3	> 6 months	China Japan
1572	Cassiopeia	-4	18 months	China Korea Europe
1604	Ophiuchus	-2.5	12 months	China Korea Europe

Let us discuss the historical records of these supernovae indivi-
dually. The two supernovae of 1572 and 1604 appeared with the emergence
of the embryo of modern astronomy and were therefore observed quite

74

precisely and carefully, by Tycho and Kepler. Figure 3.2 contains light curves of supernovae, i.e., the evolution of their luminosities, one of which is for the supernova that appeared in galaxy IC4182 in 1937. The three curves are similar in shape, so it follows that they belong to the same type of phenomenon.

The supernova of 1408 was only recorded in China and Japan. Its luminosity was greater than that of Vega. The supernova of 1181 was also only recorded in China and Japan. It had a smaller luminosity and the maximum of visual magnitude was about zero, similar to Vega. It is difficult to analyse it further as there are few records concerning the evolution of its luminosity.

The supernova of 1054 is the most famous historically because it has played a very important role in modern astronomy, especially for relativistic astrophysics. It was recorded in China and Japan, but there are more records from China and these are of interest.

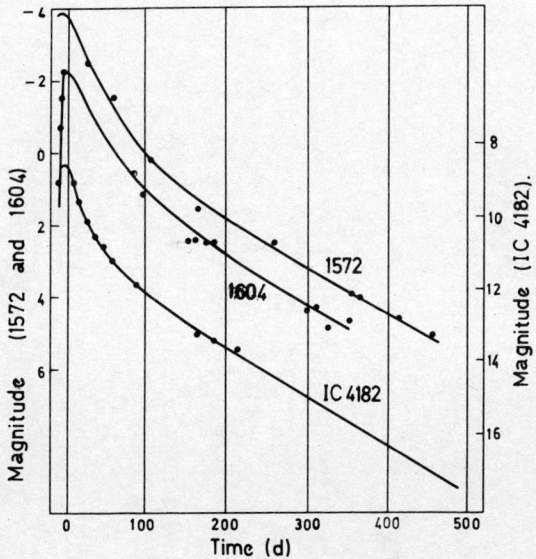

Fig. 3.2 Light curves of three supernovae

Let us cite some of the famous paragraphs:

First year of the Jaiyou reign, third month, day of Xinwei: the Director of the Astronomical Bureau reported that since the fifth month of the first year of the Zhiho reign, a guest star had appeared in the morning in the east guarding Tianguan, and it has now vanished.

Zi-Zhi-Tung-Jian

First year of the Zhiho reign, fifth month, day of Jichou: a guest star appeared approximately several inches to the south-east of Tianguan.   It has gradually vanished over more than a year.

Sung-Shi

Since the fifth month of the first year of the Zhiho reign, a guest star had appeared in the morning in the east, guarding Tianguan. It was visible in the daytime, like Venus.   It had pointed rays on all sides and its colour was reddish-white.   Altogether, it was visible for 23 days.

Sung-Hui-Yao

From these records, we know certain important facts which may be roughly divided into the following aspects:

1. Position: the supernova was situated near Taurus (Tianguan) with an estimated error of about 2^0.

2. Colour: reddish-white.

3. Magnitude: Venus was used as a comparison to describe the luminosity of the supernova.

4. Variation of Luminosity: the supernova could be seen in the day for the first twenty-three days, and was not seen by the naked eye after twenty-two months.

5. Timing: the timing (year, month and day) of the various phenomena was given.

From the data, we can sketch the light curve, from which there is no doubt that it is a supernova of type 1. We will elaborate further on this identification and its significance in the next section.

The supernova of 1006 may have been the brightest in recorded history:

The case is similar to the semi-lunar.
"History of Song Dynasty",
Astronomical Annals.

The luminosity is the same as that of the Moon.
Arabic record, "Kitab al-Muntazam".

Indeed so bright was the supernova of 1006 that it was also recorded in Europe. This nova is perhaps the only one on record from mediaeval Europe.

Two earlier supernovae, 393 A.D. and 185 A.D., were only recorded in China. There is no record on the luminosity of the supernova of 393 A.D. (However, it was clearly recorded that the supernova could be seen with the naked eye for eight months, from which it can be estimated that its maximum luminosity is about stellar magnitude -1. Records of the supernova of 185 A.D. are found in "Literature of the Later Han". It was discovered on the 7th of December 185 A.D., and disappeared from view between the 24th of July and the 21st of August in 187 A.D., after a period of visibility of 20 months. It was situated between Centaurus α and β.

3-2 Crab Nebula

Only eight guest stars with lifetimes greater than half a year have appeared in the past two thousand years, with an average of one supernova appearing every three centuries. Since the advent of the telescope in astronomy, this type of celestial object which can be seen with the naked eye has not been observed.

What relation is there between the supernovae recorded historically and astronomy today? To answer this we need to refer to studies on nebulae.

Only several nebulae can be seen with the naked eye, of which the most famous in the northern hemisphere is the Andromeda nebula. The use of the telescope has resulted in many nebulae being discovered, one

77

of which is the so-called Crab nebula discovered first by the English physicist and amateur astronomer, Beris, in 1731. Messier rediscovered it independently in 1758 and listed it in the Messier's Catalogue as the first sighting with the name M1. It was not until 1844 when Ross studied the shape of M1 and found that it looked like a crab that it became known as the "Crab" nebula.

There are many typical spiral and elliptical nebulae in the Messier Catalogue. It was difficult to make a clear distinction between the different types of nebulae, as the distances of different nebulae from the Earth were unknown until the beginning of this century. After Hubble's success in measuring the distance of the Andromeda nebula by using the Cepheid variable, it had just been known that there were two types of nebulae with very large differences in energy and dimensions, listed in the Messier's catalogue as extragalactic and galactic nebulae. The former type (for example, the Andromeda nebula) is a giant star system compared to the latter. The most typical example of the latter is the Crab nebula, on which a great deal of study has been done.

In 1921, by comparing observations over eight years, Lampland discovered the expansive motion of the Crab nebula. Duncan also obtained this result independently. The annual expansion rate of the angular diameter of the entire nebula is about 0".21.

In 1928, from the present day values of the expansion and the angular diameter (~180") of the Crab nebula, Hubble estimated that the nebula emerged about 900 years ago. In addition, the position of the supernova of 1054, which also appeared about 900 years ago, coincides quite well with the position of the Crab nebula. Hubble was therefore led to the idea that the Crab nebula is a product of the supernova of 1054. This argument is supported further by Duyrendak's analysis made in 1942.

This identification is very important because many important conclusions can be deduced from it. Trimble has analysed records over 30 years and found that the expansion speed of every part of the nebula is proportional to its distance from the centre of the nebula.

The expansion speed of the exterior of the nebula is now about 1500 km/s. If the nebula had expanded with a uniform speed since its birth, it would then follow that the nebula had emerged in 1140, which is 86 years later than 1054. This means that the expansion is neither uniform nor decelerated but accelerated, which is unexpected. The uniform acceleration is $\dot{v} \approx 10^{-5}$ m/s^2.

In order to maintain the accelerated motion, the following power is necessary:

$$\dot{W} = Mv\dot{v} + \frac{1}{2} v^2 \dot{M}$$

where M is mass of the nebula, $\dot{M}$ is the rate of the increase in mass due to the sweeping of the nebula across the interstellar medium. Taking $M = M_\odot$ and $v = 1.5 \times 10^5$ m/s, we have

$$\dot{W} = 3 \times 10^{38} + 5 \times 10^{38} n_H \text{ ergs s}^{-1} ,$$

where n_H is number density of neutral hydrogen in the interstellar medium around the nebula. If $n_H \approx 0.2$ cm^{-3}, then

$$\dot{W} \sim 4 \times 10^{38} \text{ ergs s}^{-1} .$$

To account for the source of this power is a difficult problem concerning the Crab nebula.

In 1942, Saade found that the Crab nebula has a network-like structure, between which is a background region. The optical radiation can be considered to be of approximately two parts: one with a line spectrum, from the network itself, while the other part, with continuous spectrum, is from the background region. The emissivity of the continuous spectrum is 90% of total optical luminosity of the nebula.

The radiation of the Crab nebula had first been considered as thermal plasma radiation. However, this model has encountered many problems.

1. If the background consisted of dense thermal plasma, the mass of the nebula would be as large as $20 \sim 50\ M_\odot$, which is

79

impossible, because the mass of a type 1 supernova is generally about 1.5 ~ 2 $M_\odot$.

2. If the plasma consisted of hydrogen, then there should be strong line-emission. However, no such line spectrum has ever been observed.

3. The network structure has line emission which means that the temperature of the structure should not be high. It is difficult to understand how the cold structure exists with the hot plasma in the same system.

4. It was found that the Crab nebula is an extremely strong radio-source, which cannot be explained by the thermal plasma model.

5. It has been a long time since Lampland found that the distribution of luminosity in the Crab nebula was evidently changing. As before, the thermal plasma model is unable to explain the luminosity variation.

In 1953, Shkolovsky suggested that the main mechanism of emission in the Crab nebula could be synchrotron radiation of electrons. The new model does not possess the problems and difficulties mentioned above and, moreover, some observational results can be explained naturally by it (such as the power spectrum of radiation). In addition, by using this model Shkolovsky predicted the existence of high polarization of the radiation. This prediction has been verified by observations. There is linear polarization of about 40% in the central region, but higher polarization (about 60%) in the exterior.

It is generally agreed nowadays that all emissions of the Crab nebula in different wavebands are produced by synchrotron radiation. The entire spectrum from radio to γ-rays is shown in Fig. 3.3. The luminosity of the entire nebula is about 1×10^{38} ergs s^{-1}, of which the radio part forms 12%.

The radio structure of the Crab nebula is roughly elliptical: the angular diameter of semi-intensity along the major axis (position

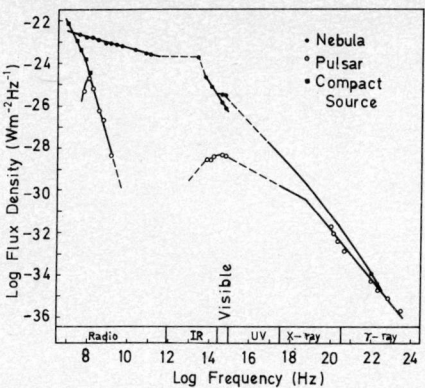

Fig. 3.3   Radiation spectrum of the Crab nebula

angle 135°) is 3'.5, and that along the minor axis is 2'.3. In the
optical range, the angular diameters of semi-intensity are 2'.5 and
1'.5 respectively. The size of the entire observable region of emission is about 7' X 5' in the radio frequency as well as in the optical range.

A relativistic electron moving in a magnetic field B would
produce synchrotron radiation, and most of the frequencies would be
near the following value:

$$\nu_m \approx 0.07 \frac{eB_{\perp}}{m_e c} \left(\frac{\varepsilon}{m_e c^2}\right)^2 \approx 1.8 \times 10^{18} B_{\perp} \varepsilon^2 \text{ Hz} \quad ,$$

where B (in units of (gauss)) is the component of the magnetic field
perpendicular to the velocity of the electron, and ε (in units of
(erg)) is the energy of the electron. If the value of the magnetic
field is taken to be 5×10^{-4} gauss and assuming that the electron
has strong emission in the radio range, then the energy of the electron
should be 2×10^{10}eV.

81

The radiant power of the electron is

$$- \frac{d\varepsilon}{dt} = \frac{2}{3} c \left(\frac{e^2}{m_e c^2} \right)^2 \left(\frac{\varepsilon}{m_e c^2} \right)^2 B_\perp^2 \approx 2.4 \times 10^{-3} B_\perp^2 \varepsilon^2 \text{ ergs s}^{-1}$$

For an electron system with the following power spectrum:

$$N(\varepsilon)d\varepsilon = K\varepsilon^{-\gamma}d\varepsilon$$

(where K and γ are constants), the specific radiant intensity in an optically thin source is

$$I_\nu = A(\gamma)K\ell B^{(\gamma+1)/2}_\nu - (\gamma-1)/2 \quad ,$$

where $A(r)$ is a function of γ, and ℓ is the dimension of the source. In the radio waveband, the power index is -0.26, so that $\gamma = 1.52$. For a magnetic field of 5×10^{-4} gauss, in order to produce the radiant intensity observed, the number density of electrons should be 10^{-5}cm^{-3}. A number density of 10^{-8}cm^{-3} is large enough to produce the observed X-ray intensity. The corresponding total energy at this density would be 4×10^{43} ergs.

From the expressions of ε and $d\varepsilon/dt$, the lifetime τ_R of the relativistic electron can be calculated as follows:

$$\tau_R \approx \varepsilon/(- \frac{d\varepsilon}{dt}) \approx 6 \times 10^{11} B_\perp^{-3/2} \nu_m^{-1/2} \text{ s} .$$

For the radio waveband frequency $\nu_m \leq 10^{11} \text{Hz}$, we have

$$\tau_R \gtrsim 2 \times 10^{11} \text{s} \approx 6000 \text{ years,}$$

which is longer than the lifetime of the supernova itself. However, the lifetime of an electron as a source of X-rays is much shorter than the supernova's lifetime. For example, when $\nu_m \approx 10^{20} \text{Hz}$, the lifetime of an electron is only

$$\tau_R \approx 6 \times 10^6 \text{ s} ,$$

i.e. 10 weeks. Therefore, if the synchrotron model is valid, we need to study the origin of the relativistic electrons: there could be either some source which continuously injects the relativistic electrons

82

into the supernova remnant, or some mechanism which re-accelerates the electrons.

In any case, the accelerated expansion of the nebula and the injection or the re-acceleration of electrons implies some source of energy, without which it would be impossible to maintain the existence of the Crab nebula.

The Crab nebula has another interesting phenomenon. There are two stars in the nebula: one with proper motion of about $\mu_\alpha = 0".000$ and $\mu_\delta = 0".000$ is in the northern region of the nebula and the other with $\mu_\alpha = -0".019$ and $\mu_\delta = 0".000$ is more southerly. On the other hand, the entire system of the Crab nebula has proper motion with $\mu_\alpha = -0".22$ and $\mu_\delta = 0".007$, which coincides very well with the southern star. Therefore, it is possible that the southern star is contained in the Crab nebula.

If this is true, then the speed of the southern star would be as large as 150 km/s or so, which is 6 to 7 times the dispersion speed of main sequence stars. This is a very peculiar feature. It is more interesting however that we have never observed any line spectrum from the southern star, but only a pure continuous spectrum. The non-validity of the ordinary classification of spectra for the star indicates that it is not a main sequence star. What type of star is it exactly? This remained a mystery for a long time.

3-3   Theory of Cold Stars

In this section we will digress from the analysis of observations in the preceding two sections and turn to basic theoretical problems.

When all thermonuclear sources of energy are exhausted, a star will gradually cool down. What will be the end result of a "cold" star? A series of theoretical studies on this problem has been done since the thirties. The basic conclusions involve two main aspects. The first is that a star in the cooling process will collapse gravitationally, and form a superdense object called a compact star; secondly, compact stars can be divided generally into three types: white dwarfs, neutron stars and black holes.

To prepare for a discussion of these two aspects of the problem, we will first study the equilibrium and stability of a star.

The basic characteristics of a star are determined by two kinds of forces: one is the self-gravitational force which causes a star to contract and the other is pressure, which opposes the contraction due to gravity. When the pressure balances the gravitational force, an equilibrium stellar system is formed. Of course, only a stable equilibrium system can be found in nature, and unstable equilibrium would disintegrate because of small perturbations.

Consider a spherical star with pressure $P(r)$ at r. Due to the difference in pressure, a shell from r to $r + \Delta r$ would be subjected to an outward-directed force on unit area:

$$P(r) - P(r + \Delta r) = \frac{-dP(r)}{dr} \Delta r \ .$$

On the other hand, the gravitational forces attracting the shell towards the centre of the star is

$$\frac{GM(r)}{r^2} \rho(r)\Delta r \ ,$$

where $\rho(r)$ is the density of matter at r, and $M(r)$ is the mass of matter contained in the sphere of radius r, that is

$$M(r) = \int_0^r 4\pi\rho(r)r^2 dr \ . \tag{3.1}$$

Equating the pressure to the gravitational forces gives

$$\frac{dP(r)}{dr} = - \frac{GM(r)}{r^2} \rho(r) \ . \tag{3.2}$$

This is the basic equation in Newtonian theory which is obeyed by a spherical star.

The relativistic equation of a spherical star is similar to that in (3.2) and has following form:

84

$$\frac{dP}{dr} = - \frac{[GM(r) + 4\pi GPr^3/c^2][\rho + P/c^2]}{r^2[1 - 2GM(r)/c^2r]} . \tag{3.3}$$

Comparing (3.2) and (3.3), the transition from Newtonian mechanics to general relativity is equivalent to the following substitutions:

$$GM(r) \rightarrow GM(r) + 4\pi GPr^3/c^2 ,$$

$$\rho \rightarrow \rho + P/c^2$$

$$r^2 \rightarrow r^2[1 - 2GM(r)/c^2r] .$$

These three substitutions also give us simple criteria for determining whether Newtonian theory is still valid or if general relativity must be used instead. If $c \rightarrow \infty$, (3.3) obviously becomes (3.2).

In the continuing discussion below, we are not going to use the equations (3.2) or (3.3) as they are, but in their variational forms. Eq. (3.2) results from the variation of

$$M = \int_0^R 4\pi\rho(r)r^2dr$$

with respect to $\rho(r)$, under the following constraint:

$$N = \int_0^R 4\pi n(r)r^2dr = const ,$$

where R denotes the radius of the star, M and N are respectively the total mass and total number of the particles in the star. In general relativity, the above two integral relations become

$$M = \int_0^R 4\pi\rho r^2dr , \tag{3.4}$$

$$N = 4\pi \int_0^R drn(r)r^2/(1 - \frac{2GM(r)}{c^2r})^{1/2} \tag{3.5}$$

where $M(r)$ is given by (3.1).

By using the approximation of uniformity, i.e. assuming ρ and n are constants in the whole interior of the star, from (3.4), (3.5)

85

and (3.1) we can immediately obtain

$$M = \frac{4\pi}{3} \rho R^3 \quad , \tag{3.6}$$

$$N = 4\pi n \int_0^R dr r^2 / (1 - \frac{8\pi \rho G r^2}{3c^2})^{1/2}$$

$$= 2\pi n R^3 (\chi - \sin \chi \cos \chi)/\sin^3 \chi \quad , \tag{3.7}$$

where $\sin \chi = (2GM/c^2R)^{1/2}$.

In order to solve (3.6) and (3.7), we also need to know the equation of state (for a cold star, the equation of state depends on temperature)

$$\rho = \rho(n) \quad , \tag{3.8}$$

or calculate the relationship between the pressure and the density from the following thermodynamical equation:

$$\frac{P}{c^2} = n \frac{d\rho}{dn} - \rho \quad . \tag{3.9}$$

From equations (3.6), (3.7) and (3.8), we can get

$$M = M(N,n) \quad .$$

The equilibrium condition of the star is

$$(\frac{dM}{dn})_N = 0 \quad ,$$

that is, the extreme points of M with respect to n. Some of these equilibrium states are stable while others are not. The criterion of stability is

$$(\frac{d^2M}{dn^2}) > 0 \quad .$$

This means that only minimum points are stable.

For a celestial body with weak gravitational fields, i.e. when $GM/c^2R \ll 1$, (3.6) and (3.7) can be expanded as power-series of GM/c^2R

$$M = N \frac{\rho}{n} \left\{ 1 - \frac{3}{5} \frac{GM}{c^2R} - \frac{99}{350} (\frac{GM}{c^2R})^2 + \ldots \right\} \quad , \tag{3.10}$$

86

$$R = \left(\frac{3N}{4\pi n}\right)^{1/3} \left\{1 - \frac{1}{5}\left(\frac{GM}{c^2 R}\right) - \frac{47}{250}\left(\frac{GM}{c^2 R}\right)^2 + \ldots\right\} , \qquad (3.11)$$

with zeroth-order solutions:

$$M^{(0)} = N\frac{\rho}{n} ,$$

$$R^{(0)} = \left(\frac{3N}{4\pi n}\right)^{1/3} .$$

The zeroth-order results are equivalent to solutions where gravity is completely ignored.

By including the first-order term of $GM/c^2 R$ in (3.10) and (3.11), we get the solution from Newtonian theory, that is,

$$M^{(1)} = N\frac{\rho}{n}\left\{1 - \frac{3}{5}\frac{GM^{(0)}}{c^2 R^{(0)}}\right\}$$

$$= N\frac{\rho}{n} - \frac{3}{5}\left(\frac{4\pi}{3}\right)^{1/3}\frac{GN^{5/3}}{c^2}\left(\frac{\rho}{n}\right)^2 n^{1/3} . \qquad (3.12)$$

The second-order solution is the post-Newtonian approximation. It is not difficult, in principle, to solve the problem order by order.

In many cases the equation of state has the following standard forms:

$$\rho = nm + Kn^\gamma \qquad (3.13)$$

where m is the mass of the particle, K and γ are constants. The first term in (3.13) denotes the contribution of the rest mass to the mass density, and the second term denotes that of the kinetic energy or interaction energy. Generally, the second term is far smaller than the first term, which means that the equation of state itself is non-relativistic. In this case, the above equation is the polytropic equation of state and from (3.9) the following relation can be deduced:

$$\frac{P}{c^2} = K'\rho^\gamma ,$$

where $K' = K\gamma/m^\gamma$. This is the standard form of the polytropic equation of state, and γ is known as a polytropic index.

Substituting (3.13) into (3.12), we have

$$M^{(1)} = Nm + NKn^{\gamma-1} - \frac{3}{5} \left(\frac{4\pi}{3}\right)^{1/3} \frac{GN^{5/3}}{c^2} m^2 n^{1/3} \quad,$$

where we have used the condition that the second term is much smaller than the first in (3.13). Using this form of M to obtain derivatives with respect to n for equilibrium and stable criteria stated previously we can immediately obtain the following conclusions: the case of $\gamma > 4/3$ corresponds to stable equilibrium, $\gamma < 4/3$ to unstable equilibrium, while $\gamma = 4/3$ is the critical case for which the relation between $M^{(1)}$ and n is

$$M^{(1)} = Nm + NKn^{1/3} - \frac{3}{5} \left(\frac{4\pi}{3}\right)^{1/3} \frac{GN^{5/3}}{c^2} m^2 n^{1/3}$$

$$= Nm + [NK - \frac{3}{5} \left(\frac{4\pi}{3}\right)^{1/3} \frac{GN^{5/3}}{c^2} m^2] n^{1/3} \quad.$$

It is obvious that in the above case $M^{(1)}$ is a monotonic function of n, and there are neither extreme values nor equilibrium solutions. However, there would be stochastic equilibrium when the following condition holds:

$$NK = \frac{3}{5} \left(\frac{4\pi}{3}\right)^{1/3} \frac{GN^{5/3}}{c^2} m^2 \quad.$$

Since $M^{(1)} \simeq Nm$, N from the above condition in $M^{(1)}$ gives

$$M^{(1)} = \left(\frac{5}{3}\right)^{3/2} \left(\frac{3}{4\pi}\right)^{1/2} \left(\frac{c^2}{G}\right)^{3/2} \frac{K^{3/2}}{m^2} \quad, \tag{3.14}$$

which is the mass of the star when $\gamma = \frac{4}{3}$.

Now we use the general conclusion mentioned above to discuss white dwarf stars, which maintain themselves against gravitational collapse by the pressure of degenerate electrons. If the model of the ideal degenerate Fermi gas is used to describe a system of electrons, then the relation between the Fermi momentum P_F of the electronic gas and the number density n of the electrons is

$$P_F = (3\pi^2 n)^{1/3} \hbar \quad, \tag{3.15}$$

and the equation of state of the electrons-nuclei system can be written approximately as

$$\rho = nm_N\mu + \frac{1}{c^2}\, n\left[\int_0^{P_F}(p^2c^2 + m_e^2c^4)^{1/2}p^2dp \bigg/ \int_0^{P_F}p^2dp\right] \quad , \qquad (3.16)$$

where m_N and m_e are the masses of nucleus and electron respectively, and μ is the ratio of the number of nuclei to electrons.

Using (3.16), (3.9) can also be written as

$$p = 6.01 \times 10^{22}f(x)$$

where $f(x) = x(2x^2 - 3)(x^2 + 1)^{1/2} + 3\sinh^{-1}x$,

$$x = \frac{P_F}{m_e c} .$$

If the number density n is not large, (3.15) shows that $P_F \ll m_e c$. Thus,

$$\rho \approx nm_N\mu + \frac{3(3\pi^2)^{2/3}}{10}\frac{\hbar^2}{m_e c^2}n^{5/3} \quad .$$

This is the polytropic equation for $\gamma = 5/3$, so it has at least a stable solution.

For this solution, as the mass of the star increases the number density n of electrons increases, that is, the Fermi momentum P_F also increases. When $P_F \gg m_e c$, the equation of state is approximately

$$\rho \approx nm_N\mu + \frac{3}{4}(3\pi^2)^{1/2}\frac{\hbar}{c}n^{4/3}$$

which is the polytropic equation for $\gamma = 4/3$. From formula (3.14), we know that the mass of a star in this critical case is

$$M^{(1)} = (\tfrac{5}{4})^{3/2}(\tfrac{3}{4\pi})^{1/2}(3\pi^2)^{3/4}\frac{c^{3/2}\hbar^{3/2}}{G^{3/2}m_N^2\mu^2} \approx 7.1\mu^{-2}M_\odot \quad , \qquad (3.17)$$

If the mass of the star increases further, n would increase and there would no longer be stable equilibrium. Therefore, $M^{(1)}$ given above is the upper limit of the mass of a star that can be supported by the pressure of degenerate electrons.

Equation (3.17) is obtained under the approximation of uniformity. The precise result is $M_{max}^{(1)} = 5.9\mu^{-2}M_\odot$, which is called the Chandrasekhar limit. It can be seen from this that the approximation of uniformity is quite good.

The density of matter in white dwarf stars can be estimated easily. When P_F is of the order of $m_e c$, the number density of electrons is

$$n \approx \frac{1}{3\pi^2}\left(\frac{m_e c}{\hbar}\right)^3 \approx 10^{30} \ cm^{-3} \quad .$$

Thus the mass density is

$$\rho = nm_N\mu = \frac{1}{3\pi^2}\left(\frac{m_e c}{\hbar}\right)^3 m_N\mu \approx 10^6\mu \ g \ cm^{-3} \quad .$$

From formula (3.11), the radius of the star is approximately

$$R = \left(\frac{3N}{4\pi n}\right)^{1/3} = \left(\frac{3M}{4\pi\rho}\right)^{1/3} \approx 10^4 \ km,$$

which is comparable to the radius of the Earth.

3-4 Neutron Stars

In 1932, Landau suggested the possibility of the existence of compact stars composed of neutrons not long after the discovery of the neutron. Baade and Zwicky also independently advanced the concept of neutron stars later, in 1934.

The basic characteristic of neutron stars is that the gravitational force is balanced by the pressure of degenerate neutrons. We know from the discussion of white dwarfs that the pressure of degenerate electrons is dominant when the density of matter is about $10^6 g \ cm^{-3}$. If the density of matter increases further, the Fermi momentum of electrons also increases, producing the following reaction:

$$p + e^- \rightarrow n + \nu_e$$

$$(Z,A) + e^- \rightarrow (Z-1,A) + \nu_e \quad,$$

the consequences of which is the neutronization of the matter. After that, the pressure of degenerate neutrons becomes dominant, and the star consists mostly of a degenerate neutron gas.

If the state of matter after neutronization is simplified as an ideal degenerate neutron gas, then the problem of the structure of neutron stars is the same as that of white dwarfs. By replacing m_e in the last section by m_N and taking $\mu = 1$, one could get roughly the results for neutron stars.

The formula (3.17) does not contain m_e, so the maximum mass of neutron stars is that of white dwarfs, i.e. several solar masses.

The Fermi momentum of neutrons is $P_F \approx m_N c$. Thus, when the number density of neutrons equals to

$$n = \frac{1}{3\pi^2}\left(\frac{m_N c}{\hbar}\right)^3 \approx 10^{31}\,\text{cm}^{-3} \quad,$$

the mass density would be

$$\rho = n m_N \approx 10^{15}\,\text{g cm}^{-3}$$

and the order of magnitude of the radius is

$$R = \left(\frac{3M}{4\pi\rho}\right)^{1/3} \approx 10\ \text{km} \quad.$$

It can be seen from this that neutron stars are much smaller and more compact.

Two approximations have been applied in the above rough estimates: one is the uniform-density approximation and the other is the ideal gas approximation. By numerically integrating eq. (3.3), one can avoid the first approximation. For example, in 1939, Oppenheimer and Volkoff, using the model of an ideal degenerate neutron gas, integrated (3.3) numerically to establish the first quantitative model of neutron stars, obtaining Fig. 3.4. The abscissa is $\tan^{-1} t_0$, $t_0 = 4\ \log(a+1\ \sqrt{1+a^2})$,

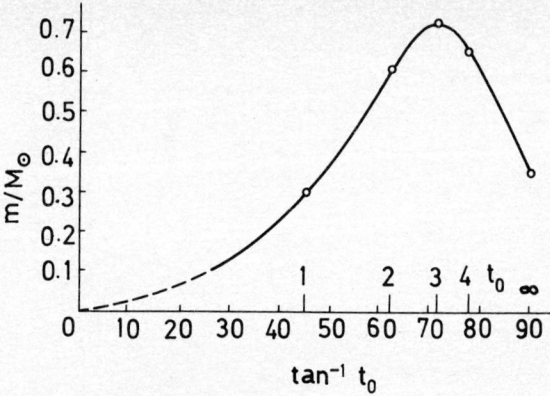

Fig. 3.4 Relation between the mass (in units of
solar mass) of neutron stars and the parameter corresponding to the central density

$a = P_F/m_N c$, where P_F is the Fermi momentum corresponding to the central density of stars; and the vertical coordinate expresses the mass
of equilibrium stars. There is a maximum of about $M = 0.7\ M_\odot$ in the
curve. It is impossible for a star with mass larger than $0.7\ M_\odot$ to
form a stable neutron star. This critical mass is generally called the
Oppenheimer-Volkoff limit.

It is possible, by using some equation of state closer to the
real case, to integrate eq. (3.3). Till today, there is no consistent
result on the equation of state for $p > 10^{14}\,g\,cm^{-3}$ mainly because of
the incompleteness of our knowledge about the interactions between nuclei. Fig. 3.5 shows the relations, obtained from different equations
of state, between the central densities and the equilibrium masses.
The upper limit for neutron stars is in the range between $0.7\ M_\odot$ and
$2.5\ M_\odot$.

92

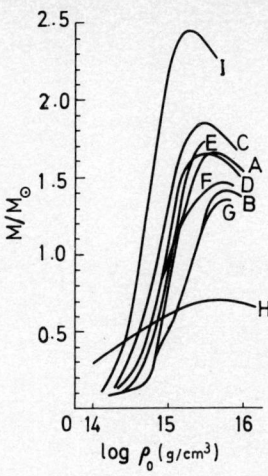

Fig. 3.5 Relations between the central density and
equilibrium mass of neutron stars calculated
by different equations of state

The density of a real neutron star varies from its centre to its
surface, so that the states are very complicated. A possible structure
is shown graphically in Fig. 3.6.

A neutron star has a solid outer shell, with a depth of about
1 km, composed of a lattice structure of nuclei and degenerate free
electrons. The density in the outer shell is about $10^6 \mathrm{g\,cm^{-3}}$, where
the major type of nucleus is $^{56}\mathrm{Fe}$. As one goes deeper starting from
the surface, one would find that the density of matter and the Fermi
momentum of electrons would gradually increase, and after entering the
neutronization region the nuclei would have more and more neutrons,
becoming neutron-rich nuclei. When the mass density is about
$4.3 \times 10^{11} \mathrm{g\,cm^{-3}}$, free neutrons begin to form, and the matter in this
domain is composed of nuclei which form the lattice, free electrons and
free neutrons. When the density rises to about $\sim 10^{14} \mathrm{g\,cm^{-3}}$, the nuclei would disintegrate completely and a kind of fluid would be formed

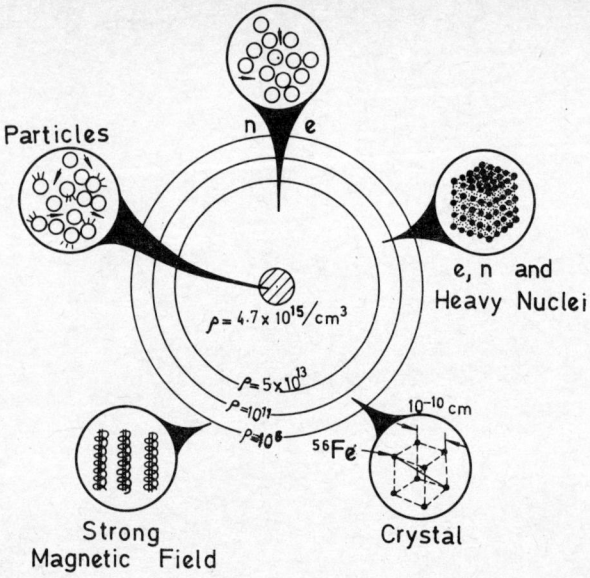

Fig. 3.6 Possible internal structure of a neutron star

in which there would be few protons, electrons or muons. This is the neutron fluid region of neutron stars.

In the neutron fluid region the neutrons form a superfluid, the protons are in a superconductive state, and the electrons are normal. This region continues to a density of the order of $10^{15} \mathrm{g\,cm^{-3}}$.

For the case when the mass density exceeds $10^{15} \mathrm{g\,cm^{-3}}$, different conclusions have emerged about the composition and the equation of state. There are three prevalent viewpoints.

1. Hyperon fluid

After the neutron fluid region, the mass density increases further, and the chemical potentials μ_n and μ_e would increase so much that the following process would occur:

$$e^- + n \to \Sigma^- + \nu_e \ .$$

94

The inverse process would be forbidden because the energy of Σ^- is less than $(\mu_n + \mu_e)$ and Σ^- is stable. Analogously, with the further increase in the density, other hyperons (like Σ^0, Σ^+, Λ, Δ and so on) could also emerge successively, and hence form a hyperon fluid.

2. Solid core of neutrons

Under the condition of high density, the neutrons may freeze, so that a solid core of neutrons would be formed in the interior of a neutron star. However, the calculated results on the critical density of the freezing point are quite different. There are also some results indicating the non-possibility of such a freeze.

3. π condensation

If the strong interaction between π mesons and other matter is neglected, then, when $(\mu_n - \mu_p)$ exceeds the rest mass of the π^- meson $M_\pi = 139.6$ MeV, π^- mesons would be formed by the process $n \rightarrow p + \pi^-$. When the density ρ is close to the nuclear density, $(\mu_n - \mu_p) \sim 100$ MeV. Therefore, it would be expected that π^- mesons are produced as long as the density is only a little higher than the nuclear density. Since π^- is a 'boson', it is possible to form a π^- condensation phase as in general Bose-Einstein condensation. However, the above conclusion is still uncertain due to the influence of strong interactions.

In 1974, C.D. Lee advanced the theory of abnormal nuclear state which suggested the possibility of the transformation of a normal nuclear state into an abnormal state when the number density of nucleons is larger than a certain critical value n_c. Since it is only a little larger than the nuclear density, the condition for the transformation may already exist in neutron stars. In other words, there may be not only neutron stars composed of neutrons in the normal state but also the so-called degenerate abnormal neutron stars composed of neutrons in the abnormal state.

One of the most interesting conclusions about the abnormal neutron stars is the possibility of the existence of "metastable" compact stars. Because both the abnormal and normal states of neutrons are

95

actually two phases of a general phase transition, there could also be "over-cooling" and "over-heating" metastable states. Therefore, it is possible that there could be metastable states in celestial bodies. Detailed calculations show that for $M < 0.8 \, M_\odot$, the normal neutron stars would be stable and abnormal neutron stars metastable; conversely when $M > 0.8 \, M_\odot$, the normal stars would be stable and abnormal stars metastable.

3-5   The Discovery of Pulsars

Before the establishing of the theory of degenerate electron stars, it had already been discovered that there was an unusual type of star called white dwarfs with masses similar to ordinary stars, possessing very small luminosities and quite high temperatures as could be deduced from their spectra.   Figure 3.7 is one of the earliest of

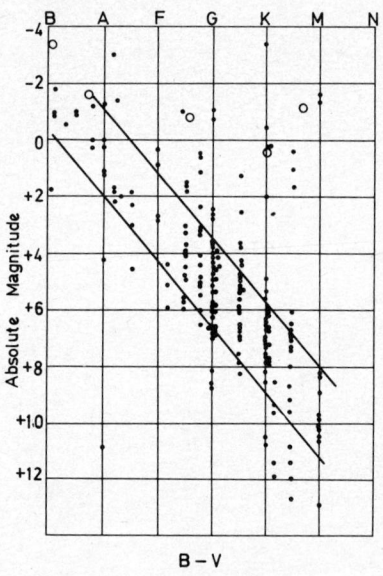

Fig. 3.7   The HR (Hertzsprung-Russell) diagram published by Russell in 1913

so-called HR (Hertzsprung-Russell) diagrams published by Russell in 1913: the vertical axis expresses the absolute magnitude, i.e. $M = -2.5 \log L + \text{constant}$, and the abscissa expresses the surface temperature decreasing from left to right. The majority of stars which are found on the oblique belt are the main sequence stars. Only one star in the figure deviates far from the main sequence and situated in the lower left region of the diagram, namely, with high temperature and low luminosity, is the white dwarf 40 Eridani B. The companion of Sirius is another white dwarf. White dwarfs are in fact the previously discussed degenerate electron stars. The development of the theory of degenerate electron stars has explained very well many of the observational data concerning white dwarfs.

The story for neutron stars would read quite differently. The theory of neutron stars was first established even before anyone knew which star was a possible candidate or how to find one, and there were even doubts as to their existence. In 1934, Baade and Zwicky made certain conjectures on this problem in a short paper which states as follows:

"Supernovae flare up in every stellar system (nebula) once in several centuries. The life time of a supernova is about twenty days and its absolute brightness at maximum may be as high as $M_{vis} = -14^m$. The visible radiation L_v of a supernova is about 10^8 times the radiation of our Sun, that is, $L_v = 3.78 \cdot 10^{48} \text{ergs s}^{-1}$. Calculations indicate that the total radiation, visible and invisible is of the order $L_t = 10^7 L_v = 3.78 \times 10^{48} \text{ergs s}^{-1}$. The supernova therefore emits during its life a total energy $E_t \geqslant 10^5 L_t = 3.78 \times 10^{53} \text{ergs}$. If supernovae initially are quite ordinary stars of mass $M = 10^{34} \text{g}$, E_t/c^2 is the same order as M itself. In the supernova processes mass in bulk is annihilated. In addition the hypothesis suggests that cosmic rays are produced by supernovae. Assuming that in every nebula one supernova occurs every thousand years, the intensity of the cosmic ray to be observed on the Earth should be of the order $\sigma = 2 \times 10^{-3} \text{ergs cm}^{-2} \text{s}^{-1}$. The observational values are about

$\sigma = 3 \times 10^{-3}$ergs cm^{-2}s^{-1}. With all reserve we advance the view that supernovae represent the transitions from ordinary stars into neutron stars, which in their final stages consist of extremely closely packed neutrons."

The important contribution of this paper is that the two different studies, the observational study concerning the supernova of 1054 with the Crab nebula and the theoretical study on cold stars, are related to each other. However, there was no observational evidence for such a connection until the discovery of the pulsar in 1967, which was a step of crucial importance in relating the two.

In 1967, Hewish and Bell *et al.* studied the interstellar scintillation of compact radio sources by employing a square array of radio antennae composed of 2048 full-wave dipoles of operating frequency 81.5 MHz, with a receiver of time resolution 0.1 s. This equipment with high accuracy and rapid recording, advantageous in discovering radio stars with rapidly variable luminances, could constantly monitor very wide regions.

With the accuracy of the telescopes of the fifties, one could have already found the pulsars. However, the existence of celestial bodies which could change rapidly was considered to be impossible (or was not considered at all) at that time, so that the time constants of receivers or recorders were generally taken as several seconds in order to smooth out irregular noise. As a result, the flux of many rapidly variable sources would have also been smoothed out, since the average flux of pulsars is smaller than the lower limit of survey flux.

Bell *et al.* first confirmed the existence of very obvious periodic signals on the 28th of November 1967, after which they made a series of excluding experiments and finally came to the conclusion that the signals were not artificial, because the parallax of the source was always smaller than 2'. This meant that the source was undoubtedly outside the solar system. In addition, the width of the pulses themselves were of about 20 ms, from which followed that the emission source could not be larger than the Earth. It was even considered at first that the signals might be from some civilization outside the Earth. However, there was no recognizable coding in the pulses and no Doppler shift in the frequency of the pulses, which meant that the source was not on a planet or a star. Several other analogous sources were then found, which proved further that pulsars were natural phenomena, some kind of star. The first pulsar was PSR1919+21. A recording observed when PSR1919+21 was discovered is shown in Fig. 3.8.

Soon after the first discovery, Steelin and Reifenstein announced the detection of a pulsar with an even shorter period of 0.0331 s in the Crab Nebula. Cocke, Disney and Taylor proved that the optical emission of this radio pulsar was also pulsatile. Lynds, Maran and Trumbo pointed out that the pulsar was exactly that star which moves together with the nebula, discussed in section 3.2.

What actually are pulsars? Firstly, the period of all pulsars is very stable, and stability over several months can be as high as 10^{-7}. This shows that the masses of the sources should be very large, because only great inertia could ensure the high stability of the period. On the other hand, the fact that the period is very short shows that the stars should be very small and compact objects; and only dwarfs and neutron stars have these characteristics.

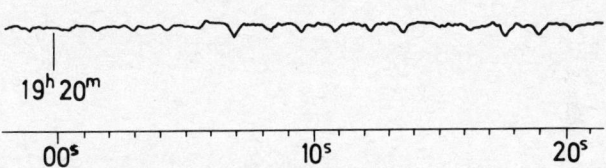

Fig. 3.8 Original recording of first observed pulsar
PSR1919+21 on 28th of November 1967

99

The following six models should be considered (see Table 6):

Table 6

Type of star	Pattern of motion		
	Radial motion	Orbital motion	Rotation
White Dwarfs	1	2	3
Neutron Stars	4	5	6

The possibilities are, for white dwarfs: (1) radial pulsations, (2) orbital motions, (3) rotations, and correspondingly (4), (5), (6) for
neutron stars, in which radial pulsations, orbital motions and rotations
are needed to produce the required periodicity of signals. Hewish has
suggested (1) and (4), Ostrike (3), Sastaw et $al.$ (5) and Gold (6).

Now let us estimate the frequency of radial pulsations. The
kinetic energy of a pulsar is

$$\frac{1}{2} M \dot{R}^2 \quad .$$

Its potential energy can be given by the function $M(n,N)$ in section
3.3. M can be expressed as a function of R and N by using the
relation connecting R to M and N. Close to the equilibrium solution, $M(R,N)$ can be expanded as follows:

$$M = M_c + \frac{1}{2} \frac{d^2 M}{dR^2}\Big|_0 (R - R_0)^2 \quad ,$$

where the subscript 0 denotes the value at the equilibrium point.
Thus, the frequency of the pulse can be expressed as

$$\nu_p = 2\pi c \left(\frac{1}{M} \frac{d^2 M}{dR^2}\right)^{1/2} \quad .$$

For example, for the polytropic equation of state:

$$\nu_p = 2\pi c \left\{ 3(\gamma - 1) \left[3(\gamma - 1) - 1 \right] \frac{N_o k \, n_o^{\gamma-1}}{M_o R_o^2} \right\}^{1/2} .$$

It is easy to show, by estimating, that the pulsation period of white dwarfs is generally greater than 1s. Therefore, the discovery of the pulsar in the Crab nebula has ruled out model (1), because the period of this pulsar is only 0.033 s. In addition, the pulsation period of neutron stars is too small to explain the long periods of certain pulsars, so that model (4) is also inadmissible.

It is necessary, in the model of orbital motions, for the white dwarf or the neutron star to be a member of a binary system. However, even if a binary system consisted of two white dwarfs, its shortest period would still be longer than several seconds, which is greater than the period of many pulsars. For a binary system composed of two neutron stars, the period would be small enough, but its gravitational radiation would be so strong (see Chapter 5) that it would cause large energy loss and the period would shorten quickly. This is contrary to observational results where the period increases slowly.

As for the rotation mechanism, the rotation period of a white dwarf cannot be smaller than 1s, otherwise the system would disintegrate due to centrifugal forces. In order to ensure that the speed of a point in the equatorial plane of the star does not exceed the speed of light, the radius of the pulsar in the Crab nebula should be smaller than 1700 km, but only the radii of neutron stars can be smaller than this value. Models (1) to (5) have been ruled out, leaving only (6), that is, pulsars should be rotating neutron stars. Neutron stars have been found to be so.

Model (6) has been supported further by semi-quantitative analysis. If one considers that a neutron star is the result of the collapse of an ordinary star, contrasted with an exploding supernova, then, due to conservation of angular momentum in the collapsing process the relation between the rotation period P_s of the star before collapse and the period P_N of the resultant neutron star after the

101

collapse would be as follows:

$$P_N \approx (\frac{R_N}{R_S})^2 \, P_S \quad ,$$

where R_S, R_N denote the radii of the star before and after the collapse respectively. The rotation period of a star is generally about one month for a star with radius of about 10^5 km, whereas the period of the neutron star formed after a short time is about 1 ms. Therefore, an ordinary star should collapse to generally become a neutron star rotating at a high speed.

Gold predicted that if the energy source of a pulsar is the rotation of the neutron star, then its period should increase. Observational results have confirmed the prediction. For instance, the rate of change of the period of the Crab nebula is dP/dt = 13.5 ns/ year. By using this value the rate of loss of rotational energy E_R,

$$\frac{dE_R}{dt} = I\omega \, \frac{d\omega}{dt} \quad ,$$

can be calculated. Taking a group of typical values: M = 0.4 $M_\odot$, R = 20 km, moment of inertia I = 4.4 X 10^{44} g cm^{-2}, and using a period P = 33 ms, the energy loss rate is

$$\frac{dE}{dt} \approx - 2 \, X \, 10^{38} \text{ergs s}^{-1} \quad ,$$

which is equal to the energy needed for maintaining the radiation and accelerating expansion of the nebula. Thus, the problem of energy sources, which was open for a long time in the study of the Crab nebula, has finally been resolved.

Recently, research has been done in comparing the supernovae of 1054 and 1006 which can be said to have almost the same age, but whose remnants are quite different. At the centre of SN (supernova) 1054 is a short period pulsar, while no pulsar nor any indication of a point source has ever been found among the remnants of SN1006.

This difference is further strengthened by X-ray observations from the Einstein satellite, which has made it possible to look for thermal emissions from the surfaces of isolated neutron stars.

It has been found that the limit on the surface temperature of the Crab pulsar is about three million degrees. This is consistent with the prediction of cooling theory for standard neutron stars. However, the limit on the temperature for SN1006 is only about a million. A normal neutron star should not be able to cool down to such a low temperature in such a relatively young star like the SN1006. This requires a faster cooling rate which can take place if the neutron star is not a normal one but is in another more exotic state such as that due to pion condensation. In such a case, the faster cooling is caused by neutrino emission.

A neutron star should undergo a phase transition during the change from a normal state to a pion state. Therefore, the further problem is whether there is any evidence of the occurrence of a phase transition in the evolution of SN1006. The answer may also be found from historical records of Chinese astronomy. Since a phase transition will generally be accompanied by a release of energy, we might expect that a re-explosion occurs during such a phase transition.

A re-explosion of this type has indeed been recorded, where a second brightening up of SN1006 was recorded on 16th May 1016, i.e. 10 years after the original explosion, at the same position.

3-6   Basic Properties of Pulsars
All of about 400 pulsars discovered up to now have certain observational features in common, of which the most important is the periodic radio emission over a wide frequency range with a pulse profile. The radio emission of PSR0329+54, a typical pulsar, is shown in Fig. 3.9, with an observed frequency of 410 MHz and a pulse period of 0.714 s. The pulse intensity changes so drastically that sometimes the pulses disappear altogether. Even so, the periodicity of the pulse is quite constant. If the time constant of an observational instrument is set lower than 1 ms, it will be found that the pulse

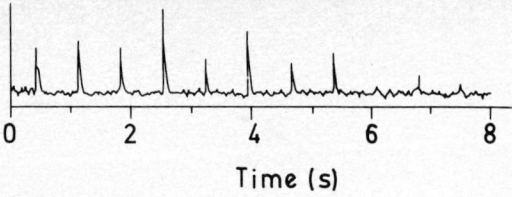

Time (s)

Fig. 3.9   Radio emission of the pulsar   PSR0329+54

has a complicated structure: every pulse is in general composed of
sub-pulses which may overlap. The width of every pulse is roughly 1 or
2 per cent of the period. If the time constant is below 10 microseconds the microstructure of the sub-pulses can now be found, where
characteristic widths are about 0.1% of the period. Since pulsars are
regarded as neutron stars, one period of pulsars is equivalent to a
rotating cycle of neutron stars and also to a change of longitude angle
of 360°. The width of a typical sub-pulse is 5° of longitude and that
of microstructure is about $0^{o}.3$.

Other than the amplitude of pulses, the phases of the sub-pulses
(i.e. profiles of the sub-pulses) also have systematic variation, which
usually synchronize with the pulse period. Even if these variations do
exist, the integrated profile of every pulsar is very stable, differing
for different pulsars. About half of all pulsars have single-peak profiles while the rest have several peaks. It is thus not rare to observe
double-peak pulses and some pulses may even have five peaks. The different kinds of typical integrated profiles are shown in Fig. 3.10.

The equivalent width of the integrated profiles of the pulses,
defined as the ratio of the energy of the pulse to the density of energy
flow of the peak, is generally equal to 10° of longitude, i.e. about 3%
of the period.

104

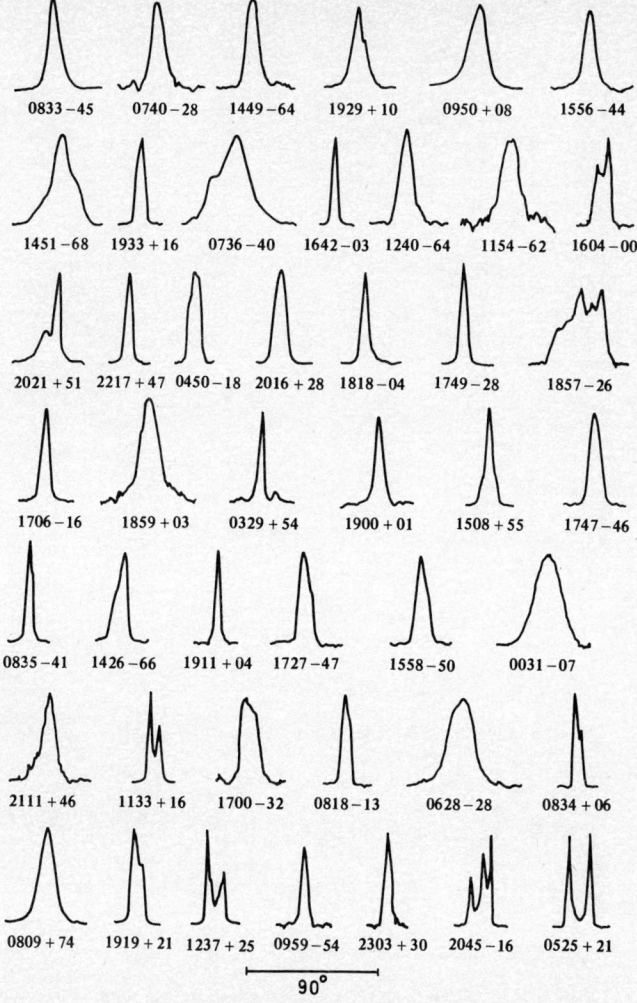

Fig. 3.10 Typical emission profiles of pulsars

The radio spectrum of pulsars is a power spectrum i.e. $S = S_o \nu^\alpha$, and is quite steep with a typical index value $\alpha = -1.5$. For many pulsars, the radio spectra are steeper at high frequencies. In Fig. 3.11, the

105

radio spectra of six pulsars are shown.

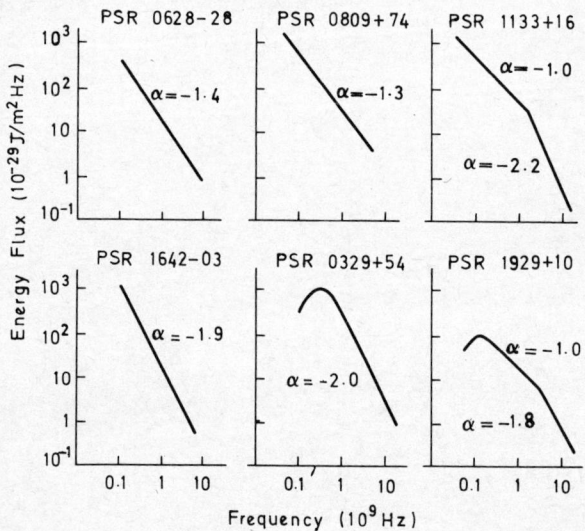

Fig. 3.11 Radio spectra of six pulsars

Fig. 3.12 is a histogram of the number of pulsars with respect to period, from which it can be seen that the majority of pulse periods are about 0.65 s. There is an obvious concave at 1 s, which may seem to indicate that there are two kinds of pulsars: one with a short period and the other with a long period.

All pulsars have increasing periods where a typical rate of increase is about 10^{-15} s per year. The stable increasing of the period indicates that the age of pulsars cannot be longer than the characteristic time $T \equiv P \dot{P}^{-1}$, a typical value being 10^7 years. For short-period pulsars, the rate of change of the period is large and their characteristic time is very small. For example, the pulsar PSR0531+21 in the Crab Nebula has $P = 422.69 \times 10^{-15}$, so that its

106

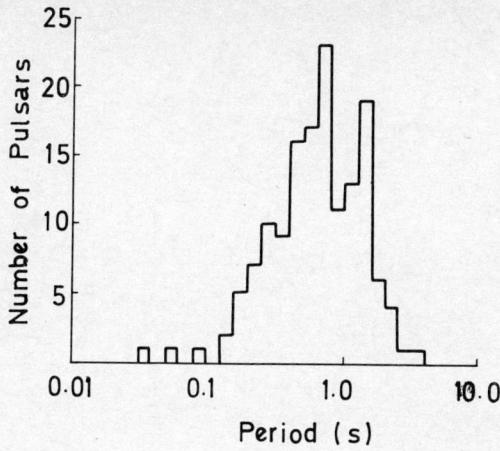

Fig. 3.12 Distribution of pulsar periods

characteristic time $T \approx 2480$ years.

The distribution of pulsars in galactic coordinates is shown in
Fig. 3.13, from which it can be seen that the majority of the pulsars
are concentrated near the galactic plane, indicating that pulsars
should belong to a galactic-disk population. In addition, as more of

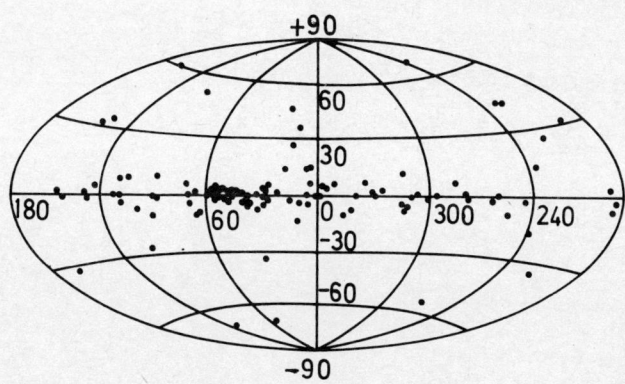

Fig. 3.13 Distribution of pulsars on the galactic plane

107

the pulsars are distributed in the direction of the galactic centre, this means that the density of the pulsars decreases with increasing galactic longitude at least near the Sun.

It is difficult to find exactly how distant pulsars are. Estimates of pulsar distances are only based on measuring the dispersion of radiations. The dispersions are proportional to the electrons per square centimeter in the intervening space along the line of sight; thus, if the number density of electrons is known, the distances can be calculated. The pulsar distances obtained in this way range from 100 pc to 2000 pc. Due to the uncertainty in distributions of electron density, the values of the distances of a few pulsars may be in error by a factor of two. However, these values of distances are statistically quite precise. The values of Z (the vertical distances to the galactic plane), for the majority of pulsars, are smaller than 300 pc. Therefore, the distribution of the pulsars is disc-like in shape. Fig. 3.14 shows the projection of pulsars on the galactic plane, where the spiral structure does not appear in the distribution of the pulsars.

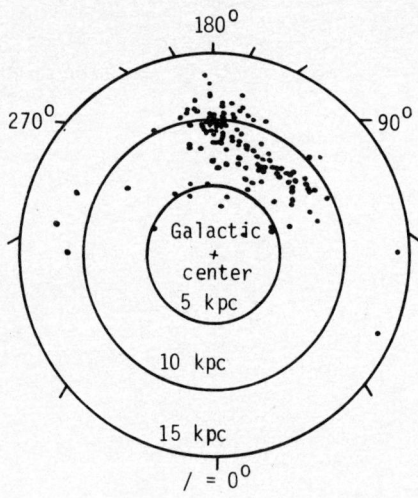

Fig. 3.14 Projection of the space distribution of pulsars on the galactic plane

How does the rotational energy of neutron stars get transformed
into the energy of radiations and high speed particles? The problem on
the transformation mechanism of energy has not yet been resolved and it
has become more acute after the discovery of γ-ray pulses. Many
radio pulsars with strong emissions of high energy γ-ray pulses have
been found by employing the satellite COS-B since 1978. The Crab
pulsar PSR0531+21 and PSR0833+45, PSR0740-28 and PSR1822-09 have
all been identified completely, their γ-ray emissions possessing some
common features. Firstly, all luminances of γ-rays are very great,
about 10^6 times those of corresponding respective radio emissions.
In other words, radio emissions is the only "negligible" part of the total emission, so that it may be more suitable for these objects to be
called γ-ray pulsars rather than radio pulsars. Secondly, all pulsars
have similar dual-peak structures in their γ-rays spectra, with phase
differences between the two peaks of about 0.4 (see Fig. 3.15). The

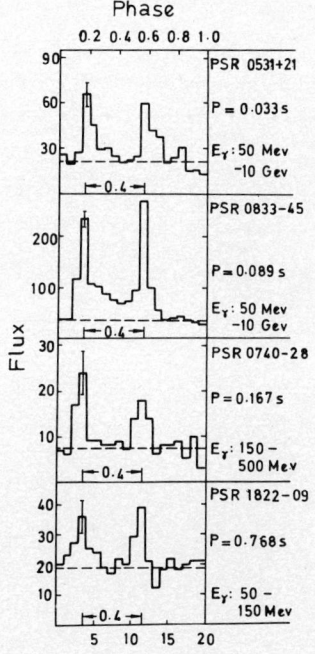

Fig. 3.15 Dual-peak structures of γ-ray pulses

109

luminosity variations of γ-rays with the periods of the pulsars are very weak, but the transformation rates of rotational energy into γ-ray energy increase with the age of the pulsars. All these properties differ greatly from those of radio emissions, and this point could cause extremely great changes in our knowledge about pulsar emissions.

REFERENCES

On supernovae:

L. Woltjer, Annual Review of Astronomy and Astrophysics <u>10</u> (172).

On the theory of neutron stars:

L. Landau, Physikalishe Zeitschrift der Sowjetunion <u>1</u>, 285 (1932).

J.R. Oppenheimer, G.M. Volkoff, Phys.Rev. <u>55</u>, 375 (1939).

A.G.W. Cameron, Annual Review of Astronomy and Astrophysics <u>8</u>, (1970).

V. Canuto, Annual Review of Astronomy and Astrophysics <u>12</u>, (1974); <u>13</u>, (1975).

L.Z. Fang, G.Z. Xie, KEXUE TONGBAO, <u>24</u>, 167 (1979).

On pulsars:

A. Hewish, Annual Review of Astronomy and Astrophysics <u>8</u>, (1970).

R.N. Manchester, J.H. Taylor, Pulsars (W.H. Freeman and Company, San Francisco, 1977).

Chapter 4

BLACK HOLES

4-1   Critical Mass

From the last chapter, we know that there are upper limits on the masses of neutron stars. Oppenheimer and Volkoff performed the first numerical calculations on the structure of neutron stars by assuming the equation of state of the degenerate free-neutron gas and by using the equilibrium eqs. (3.1) and (3.3) from general relativity, which can be rewritten as follows:

$$\frac{dM(r)}{dr} = 4\pi\rho r^2 = H(\rho,r) \quad ,$$

$$\frac{dP}{dr} = -\frac{[GM(r) + 4\pi GPr^3/c^2][\rho + P/c^2]}{r^2[1 - 2GM(r)/c^2 r]} = G(\rho,P,m,r) \quad .$$

It was pointed out in the last chapter that the upper limits on the masses depend directly on the equation of state. Within the range $\rho \leqslant 4.6 \times 10^{14} \, \text{g cm}^{-3}$ where attraction is more important than interactions between nucleons, the equations of state obtained by using different methods are the same because studies in nuclear physics have already provided us with the possibility of getting a suitable equation of state. However, above the limit $\rho > 4.6 \times 10^{14} \, \text{g cm}^{-3}$ where repulsion dominates, there is a need for knowledge about the hard core of

111

nucleons, but even up to now the physics of nuclear matter in this domain is only poorly understood. As a result, the equations of state in
this domain are usually based on an extension or some approximation
without sufficient experimental basis.

On the other hand, the masses of neutron stars are contributed
mainly by the high density regions. For instance, if the central den-
sity of neutron stars $\rho \approx 3 \times 10^{15} \, \text{g cm}^{-3}$, then only several per cent
of the total mass of a neutron star would be contributed by the domain
$\rho < 4.6 \times 10^{14} \, \text{g cm}^{-3}$. It follows that the upper limits of the masses of
neutron stars would be very strongly influenced by the uncertainty in
the equations of state. In other words, the evaluation of the upper
limits of the masses would not be based on the concrete equations of
state but only on the very general demands satisfied by the state
concerned; that is, we accept the following assumptions as starting
points:

1. The equilibrium eqs. (3.1) and (3.3) of hydrostatics in general relativity are to be obeyed by the configurations of
   the stars;

2. In the region $\rho > 4.6 \times 10^{14} \, \text{g cm}^{-3}$, we assume the state to
   have the following properties:

   (a) there is no local spontaneous instability, which is
       equivalent to

$$\frac{dP}{d\rho} \geq 0 \quad , \tag{4.1}$$

   i.e. the pressure, which is positive, is also a mono-
   tonous function of the density;

   (b) due to causality, the speed of sound will be less than
       the speed of light, namely

$$\sqrt{\frac{dP}{d\rho}} \leq c \quad ; \tag{4.2}$$

   or (4.1) and (4.2) combined together to give

$$0 \leq \frac{dP}{d\rho} \leq c^2 \quad . \tag{4.3}$$

112

Our problem is to try to evaluate the upper limits on the masses of neutron stars by using (3.1), (3.3), and (4.1) to (4.3).

Let us first analyse the problem qualitatively. If we use the polytropic equation of state $P = K\rho^{\gamma}$, the condition $dP/d\rho > 0$ means $K\gamma \geqslant 0$, and $dP/d\rho \leqslant c^2$ means

$$K\gamma\rho^{\gamma-1} \leqslant c^2 \tag{4.4}$$

We know that the larger the mass of a star the higher would be its central density, which undoubtedly indicates that condition (4.4) cannot be satisfied for a sufficiently large star unless $\gamma < 1$. On the other hand, the discussion in section 3.3 has already shown that it is impossible for $\gamma < 4/3$ to have a stable solution. This indicates that there certainly are upper limits on the masses. It will be seen from this analysis that (4.4) is a rigorous condition, and in spite of the fact that it is based on the polytropic equation of state, the conclusion is still valid for general cases.

The above general problem can be treated mathematically by the variational method, that is, by evaluating the maximum M. We write M as

$$M = \int_{\rho_0}^{\rho_c} \frac{dM(r)}{d\rho} \, d\rho + M_e(\rho_0, P_0, r_0, M_0) \tag{4.5}$$

where $\rho_0 = 4.6 \times 10^{14} \, \mathrm{g\,cm}^{-3}$, M_e is the mass in the layer $\rho < \rho_0$ where the equations of state are known and where we can carry out the integration by using the usual methods; ρ_0, P_0, r_0, M_0 denote the respective quantities in the layer $\rho = \rho_0$. The first term of (4.5) expresses the mass of the region from ρ_0 to the centre of the star where the density is ρ_c. Using the variational equation and restrictive conditions (3.1), (3.3) and (4.3) to solve the first term of (4.5), we get the result

$$M = 3.2 \, M_\odot \quad ,$$

which is the critical mass of neutron stars. All results obtained by employing definite equations of state should be smaller than this value,

113

which is therefore called the absolute critical mass. The present
upper limit

$$M = 3.17 \, M_\odot \quad ,$$

evaluated by using the equations of state of abnormal neutrons, is the
largest of several results evaluated by using definite equations of
state. It follows that there will be no other type of compact star
with masses greater than the critical value as long as the theory of
abnormal state of neutrons is correct. Any cold star more massive than
the critical value will collapse infinitely to form a black hole.

4-2   The Phenomena in a Collapse

Black holes are very strange objects with many peculiar properties. We will discuss some properties of Schwarzschild black holes in
this section. What is known as a Schwarzschild black hole is the
spacetime described by the Schwarzschild metric with special attention
given to the region $r < 2GM/c^2$.

We know that a particle may remain stationary at a certain point
(r, θ, ϕ) outside a black hole $(r > 2GM/c^2)$, but it is impossible for
a body inside the black hole to be at rest, i.e. to maintain fixed values of coordinates (r, θ, ϕ). The argument runs as follows: first we
recall that the world line of any particle must be time-like, namely,
the separation $d\tau^2$ along the world line is always positive. However,
for a particle "at rest" with

$$dr = d\theta = d\phi = 0 \quad ,$$

equation (2.3) gives

$$d\tau^2 = (1 - \frac{2GM}{c^2 r}) dt^2 \tag{4.6}$$

so that when $r < 2GM/c^2$, $d\tau^2$ is negative, which means that it is
impossible for a particle to have $\Delta r = 0$.

How do bodies move inside a black hole? To answer this we
examine the properties of light cones. There are double cones at any
event P: one points into P's past, while the other points into P's
future. The time-like geodesics from P, for which $\Delta\tau^2 > 0$, are all

114

from the past cone into the future cone.

By using the following transformation

$$\tilde{t} = t + \left(\frac{2GM}{c^3}\right) \ell n \left| \frac{rc^2}{2GM} - 1 \right| \qquad (4.7)$$

the Schwarzschild metric (2.3) becomes

$$d\tau^2 = \left(1 - \frac{2GM}{c^2 r}\right) d\tilde{t}^2 - \frac{4GM}{c^3 r} dr \, d\tilde{t}$$

$$- \left(1 + \frac{2GM}{c^2 r}\right) \frac{1}{c^2} dr^2 - \frac{1}{c^2} r^2 d\theta^2 - \frac{1}{c^2} r^2 \sin^2\theta d\phi^2 \quad . \qquad (4.8)$$

Therefore, for light propagating along the radial direction (i.e. $d\theta = d\phi = 0$), we have

$$\left(1 - \frac{2GM}{c^2 r}\right) d\tilde{t}^2 - \frac{4GM}{c^3 r} dr \, d\tilde{t} - \left(1 + \frac{2GM}{c^2 r}\right) \frac{1}{c^2} dr^2 = 0 \quad ,$$

i.e.

$$\frac{d\tilde{t}}{dr} = \begin{cases} -\dfrac{1}{c} \\[2mm] +\dfrac{1}{c}\left(\dfrac{1 + 2GM/c^2 r}{1 - 2GM/c^2 r}\right) \end{cases} . \qquad (4.9)$$

Consequently, we can draw the light cone for every even point (see Fig. 4.1). The cones which correspond to increasing coordinate time $\tilde{t}$ are future cones; those cones which correspond to decreasing $\tilde{t}$ are past

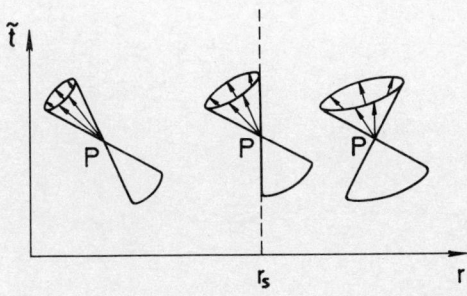

Fig. 4.1 Light cones near a black hole

115

cones. When $r > r_s$, there are some time-like lines in the future cones corresponding to increasing r, and others to decreasing r, that is, particles passing through P can travel either inwards or outwards. As we approach the black hole, the future cones tilt inwards (towards the left). When $r < r_s = 2GM/c^2$, one can only find geodesics corresponding to decreasing r in the future cones, that is, all bodies inside a black hole move towards $r = 0$. It is impossible to prevent this infinite collapse by applying forces. Any attempt to reverse time is futile.

In the standard form (2.3) of the Schwarzschild metric, when $r = 2GM/c^2$, g_{rr} becomes infinite. However, $r = 2GM/c^2$ is not an intrinsic singularity, because spacetime itself is not singular there. If we use other suitable coordinate systems, $g_{\mu\nu}$ at the so-called horizon of a black hole is not infinite. In fact, after the transformation (4.7), g_{rr} in (4.8) is no longer infinite at $r = 2GM/c^2$. This singularity which can be eliminated by the choice of suitable coordinate systems is called a coordinate singularity. The situation is similar to that in ordinary flat space using polar coordinates when $g_{\theta\theta}$ and $g_{\phi\phi}$ become infinite when r is infinite; if we use Cartesian coordinates, these singularities will disappear.

In the Schwarzschild metric (2.3), the centre at $r = 0$ is an intrinsic singularity which cannot be eliminated by any transformation of coordinates, because the spacetime curvature itself is singular there. We have

$$R^{\mu\nu\lambda\sigma} R_{\mu\nu\lambda\sigma} = \frac{12}{r^6} \left(\frac{2GM}{c^2}\right)^2 .$$

Curvature is an intrinsic geometric property and its singularity is independent of any coordinate system.

We now study the history of a body falling radially into a black hole starting from rest at infinity. Imagine an observer on the body with a clock which gives the proper time τ. First, we calculate its radial coordinate r as a function of the proper time τ. From (2.8), let

$$\left(1 - \frac{2GM}{c^2r}\right) \frac{dr}{d\tau} \equiv A \quad,$$

to be used in (2.6) to give

$$\frac{1}{2}\left[\left(\frac{dr}{d\tau}\right)^2 + r^2\left(\frac{d\phi}{d\tau}\right)^2\left(1 - \frac{2GM}{c^2r}\right)\right] - \frac{GM}{r} = \frac{c^2}{2}(A^2 - 1) \equiv E \quad.$$

Since there is no angular motion, $d\theta = d\phi = 0$ and the above equation becomes

$$\frac{1}{2}\left(\frac{dr}{d\tau}\right)^2 - \frac{GM}{r} = \frac{c^2}{2}(A^2 - 1) \equiv E \quad.$$

This is of the same form as the Newtonian equation for the conservation of mechanical energy with the first term on the left for kinetic energy and the second term for potential energy. When $r \to \infty$ and $dr/d\tau = 0$, then $E = 0$ or $A^2 = 1$, so that

$$r\left(\frac{dr}{d\tau}\right)^2 = 2GM \quad.$$

If the travelling clock is set so as to read $\tau = 0$ at the moment of collapse at $r = 0$, then the solution of this equation is

$$r = \left[B - \frac{3}{2}\tau\sqrt{2GM}\right]^{2/3} \quad.$$

If the integration constant B is made zero, then we have

$$r = \left(-\frac{3\tau}{2}\right)^{2/3}(2GM)^{1/3} \quad,$$

Thus the entire falling process of the body is from $\tau = -\infty$ (at $r = \infty$) to $\tau = 0$. It takes a time of

$$0 - \tau(r = r_s) = +\frac{2r_s}{3c}$$

for the body to move from $r_s = 2GM/c^2$ to $r = 0$. For a black hole of solar mass with $r_s = 2.95$ km, the collapse time $|\tau(r = r_s)|$ is about 10^{-5} s which means that the collapse is very rapid after crossing the boundary at r_s.

Since the light cones always point inwards for $r < r_s$ so that no signal, optical or material, can escape from region $r < r_s$ to region $r > r_s$, that is, there is no time-like geodesic from $r < r_s$ to $r > r_s$, no observer outside the black hole can see the final stages of collapse. The boundary at $r = r_s$ forms a critical surface called the horizon.

For an external observer at rest at a certain point (r,θ,ϕ) far away from the black hole, his clock gives time t. Using t in place of τ in $r(\tau)$, we can get $r(t)$ by employing (2.8), that is $(1 - 2GM/c^2r)dt/d\tau = A$. Notice that when $r \rightarrow \infty$, $t = \tau$, so that $A = 1$. Thus

$$dt = \frac{d\tau}{1 - r_s/r(\tau)} = \frac{r(\frac{d\tau}{dr})dr}{r - r_s} = - \frac{r^{3/2}dr}{\sqrt{2GM}\,(r - r_s)} \ .$$

Integrating this equation, we have a solution

$$t = \frac{r_s}{c}\left[- \frac{2}{3}\left(\frac{r}{r_s}\right)^{3/2} - 2\left(\frac{r}{r_s}\right)^{1/2} + \ln\left|\frac{1 + (r/r_s)^{1/2}}{1 - (r/r_s)^{1/2}}\right| \right] ,$$

corresponding to the world line sketched in Fig. 4.2. We see that t becomes infinite as the body approaches r_s which means that an

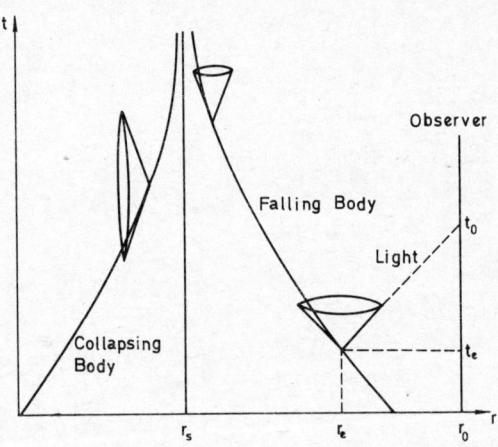

Fig. 4.2 World line near a black hole

118

observer at infinity takes infinite time to observe the arrival of the body at r_s, and when nearing r_s it appears to be frozen at that point. It is for this reason that black holes are also known as "frozen" stars.

We will next discuss the properties of the light emitted at various stages in a collapse. Light emitted at the event (t_e, r_e) travels radially outwards along a null geodesic to reach a distant observer at the event (t_o, r_o) (Fig. 4.2). Along such a world line, r and t are related by (2.3) with $d\tau = d\theta = d\phi = 0$, that is:

$$c^2(1 - \frac{2GM}{c^2 r})dt^2 - (1 - \frac{2GM}{c^2 r})^{-1}dr^2 = 0 \quad .$$

Therefore

$$t_o - t_e = \frac{1}{c} \int_{r_e}^{r_o} \frac{dr}{1 - 2GM/c^2 r}$$

$$= \frac{r_o - r_e}{c} + \frac{2GM}{c^3} \ln\left(\frac{r_o - r_s}{r_e - r_s}\right) \quad . \tag{4.10}$$

It follows that to travel from $r_e = r_s$ to r_o, light would require an infinite amount of time so that it would never be observed by an external observer.

Since light has to climb out of a strong gravitational field, it will be redshifted if received at r_o, and it is of interest to calculate this redshift. Consider two signals emitted by an atom at r_e (assume that they correspond respectively to the two ends, initial and latter, of an entire wavelength) with time coordinates t_e and $t_e + \Delta t_e$ and received later at t_o and $t_o + \Delta t_o$ respectively. Since the Schwarzschild metric $g_{\mu\nu}$ is independent of t, then $\Delta t_o = \Delta t_e$ (Fig. 4.3). Between the emission events, the proper time is, from (2.3),

$$\Delta\tau_e \equiv \frac{1}{\nu_e} = \frac{\lambda_e}{c} = \Delta t_e \sqrt{1 - \frac{2GM}{c^2 r_e}} \quad ,$$

119

where ν_e and λ_e are the frequency and wavelength as measured in its rest frame. For the observer, analogous relations hold:

$$\Delta\tau_0 \equiv \frac{1}{\nu_0} = \frac{\lambda_0}{c} = \Delta t_0 \sqrt{1 - \frac{2GM}{c^2 r}} \quad .$$

Thus the formula for the redshift is

$$Z = \frac{\lambda_0 - \lambda_e}{\lambda_e} = \sqrt{\frac{1 - 2GM/c^2 r_0}{1 - 2GM/c^2 r_e}} - 1 = \frac{r_e(r_0 - r_s)}{r_0(r_e - r_s)} - 1 \quad .$$

As $r_e \to r_s$, Z becomes infinite, so that the light emitted by a star just as it collapses into a black hole is infinitely redshifted and hence unobservable. If $r_e \gg r_s$ and $r_0 \gg r_s$, the redshift reduces to our earlier approximation results in section 2.1.

The sense of the redshift is such that the stationary clock at r_e differs in time from that at rest at r_0. Therefore, not only would the frequency of the light emitted by a collapsing star be redshifted but the number of the photons emitted per unit time would also decrease by the same factor. Thus, the initial luminosity L_1 of a star will become, because of the redshift,

$$L(t_0) = \frac{L_1}{(1 + Z)^2} = L_1 \frac{(r_e - r_s)r_0}{(r_0 - r_s)r_e} \quad .$$

Equation 4.10 relates t_0 to r_e, simplifying when $r_e \approx r_s$ to

$$t_0 \approx (\frac{r_s}{c}) \ln \frac{(r_0 - r_s)}{(r_e - r_s)} \quad .$$

Therefore, the luminosity of a collapsing star before contracting to less than r_s changes with t_0 as

$$L(t_0) \approx L_1 (\frac{r_0}{r_e}) \exp(-ct_0/r_s) \quad ,$$

which shows that the power received diminishes exponentially with t_0. Using the apparent magnitude m to denote the luminosity, we have

$$m = 2.5 \log \left(\frac{L_1}{L(t_0)}\right) + \text{const} = 1.14 \frac{ct_0}{r_s} + \text{const}.$$

This indicates that the luminosity of a star of solar mass changes very rapidly during its collapse to its gravitational radius r_s, for example in only about 10^{-5} s for an increase in apparent magnitude of 1.

4-3   The Types of Black Holes

Any collapsing star more massive than the critical mass must reduce itself into a black hole. Despite the known fact that, in general, stars may differ greatly from each other in their properties such as mass, angular momentum, charge, magnetic momentum and chemical compositions, there are actually only a few types of black holes that are known to have evolved from this range of different stars.

There is a theorem on this problem which states that stationary black holes are only of one simple type, the metric of which is

$$d\tau^2 = \frac{\Delta}{\rho^2} [dt - \frac{1}{c} a \sin^2\theta d\phi]^2 - \frac{\sin^2\theta}{\rho^2} [\frac{1}{c^2}(r^2 + a^2)d\phi - adt]^2$$

$$- \frac{\rho^2}{c^2\Delta^2} dr^2 - \frac{1}{c^2} \rho^2 d\theta^2 \quad , \tag{4.11}$$

where

$$\Delta \equiv r^2 - 2 \frac{GM}{c^2} r + a^2 + \frac{GQ^2}{c^4}$$

$$\rho \equiv r^2 + a^2 \cos^2\theta$$

$$a = \frac{L}{Mc}$$

where M, L and Q denote mass, angular momentum and charge of a black hole.

We will first discuss special cases of the metric (4.11). When $L = Q = 0$ but $M \neq 0$, (4.11) becomes the Schwarzschild metric (2.3)

121

that is
$$d\tau^2 = \left(1 - \frac{2GM}{c^2r}\right)dt^2 - \frac{1}{c^2}\left[\left(1 - \frac{2GM}{c^2r}\right)^{-1}dr^2 + r^2d\theta^2 + r^2\sin^2\theta d\phi^2\right].$$
$$\text{(4.12)}$$

When $L = 0$ but M and Q do not vanish, (4.11) becomes
$$d\tau^2 = \left(1 - \frac{2GM}{c^2r} + \frac{GQ^2}{c^4r^2}\right)dt^2$$
$$- \frac{1}{c^2}\left[\left(1 - \frac{2GM}{c^2r} + \frac{GQ^2}{c^4r^2}\right)^{-1}dr^2 + r^2d\theta^2 + r^2\sin^2\theta d\phi^2\right] \quad, \text{(4.13)}$$

which is known as the Reissner-Nordstrom metric. When $Q = 0$ but M and L do not vanish, (4.11) becomes

$$d\tau^2 = \left(1 - \frac{2GM}{c^2\rho^2}\right)dt^2 - \frac{2}{c}\left(\frac{2GMr}{c^2\rho^2}\right)a\,\sin^2\theta\,dt\,d\phi$$
$$- \frac{1}{c^2}\frac{\rho^2}{\Delta'}dr^2 - \frac{1}{c^2}d\theta^2 - \frac{1}{c^2}\frac{\Lambda}{\rho^2}\sin^2\theta d\phi^2 \quad, \qquad \text{(4.14)}$$

where
$$\Delta' = r^2 + a^2 - \frac{2GMr}{c^2} \quad,$$
$$\Lambda = (r^2 + a^2)^2 - a^2\sin^2\theta \quad,$$

which is the Kerr metric. This theorem can be summarized in Table 7, which shows the types of the stationary black holes.

Table 7. Types of black holes

Type	Characteristics	Metric
Schwarzschild	$M{\neq}0$, $L{=}Q{=}0$	(4.12)
Kerr	$M{\neq}0$, $L{\neq}0$, $Q{=}0$	(4.14)
Reissner-Nordstrom	$M{\neq}0$, $Q{\neq}0$, $L{=}0$	(4.13)
Kerr-Newman	$M{\neq}0$, $L{\neq}0$, $Q{\neq}0$	(4.11)

We do not intend to prove the theorem but only to analyse its meaning. The theorem indicates that black holes are very simple objects, and can be completely specified by three physical parameters (i.e. M,L,Q), which is to say, black holes interact with the external environment only through the effects caused by mass, charge and angular momentum; no further interaction comes into play. The conservation of mass, angular momentum and electric charge still holds for black-hole physics, but conservation laws concerning, for example, the strong and weak interactions are no longer effective because there are no strong or weak interactions between a black hole and external matter. For instance, the conservation of both lepton number and baryon number will be transcendental for a black hole. A collapsing process therefore causes complicated stars, possessing many different physical proper-ties, to transcend into simple objects that may be specified by only three parameters. This simplification has led some to say that 'a black hole has no hair'.

4-4   Horizons and Emissions from Black Holes

The most distinct feature of a black hole is the existence of a closed horizon. For a Schwarzschild black hole, the surface $r = 2GM/c^2$ constitutes the horizon. The fundamental property of a horizon is that an observer outside the horizon cannot obtain any information about the region within it.

As discussed in the last section, the most general stationary black holes are the Kerr-Newman black holes with horizons at

$$r_+ = \frac{GM}{c^2} + (\frac{G^2M^2}{c^4} - \frac{G^2Q^2}{c^8} - a^2)^{1/2} \quad , \qquad (4.15)$$

and the area of the horizon is

$$A = 4\pi(r_+^2 + a^2) \quad . \qquad (4.16)$$

When $L = Q = 0$, the above formula simplifies to

$$A = 4\pi r_s^2 \quad .$$

This form is to be expected, because the area of a surface $r = r_o$ in the standard Schwarzschild metric should also be equal to $4\pi r_o^2$.

Particles and photons can fall inwards through the horizon but none can ever emerge outwards through it. Thus, there must be irreversibility in the evolution of black holes, but how does one formulate irreversibility? Hawking has proved a theorem which states that the area of the horizon during the evolution process of a black hole can never decrease. This irreducibility of area can be expressed as

$$dA \geqslant 0 \ . \tag{4.17}$$

We define the mass m_{ir} as

$$m_{ir}^2 = \frac{1}{16\pi} A \frac{c^4}{G^2} \ . \tag{4.18}$$

The mass m_{ir} can also never decrease, because of the irreducible area, and is therefore an irreducible mass. From the definition (4.18), we can rewrite (4.16) as

$$M^2 = \left(m_{ir} + \frac{Q^2}{4Gm_{ir}} \right)^2 + \frac{c^2}{G^2} \frac{L^2}{m_{ir}^2} \tag{4.19}$$

which in physical terms states that the mass M (or the total energy) of a black hole consists of its irreducible mass m_{ir}, its rotational energy (the term with L^2) and its Coulomb energy (the term with Q^2). The total mass M, unlike m_{ir}, can in practice increase or decrease, which means that only processes concerning rotational and Coulomb energy are reversible.

For example, we can extract rotational energy from a Kerr black hole. By direct differentiation of (4.19), we have

$$dM = \frac{K}{8\pi} dA + \Omega dL + V dQ \tag{4.20}$$

where

$$K = \frac{1}{(r_+^2 + a^2)} \left[m_{ir}^2 - \frac{Q^2}{G} - \frac{c^2 a^2}{G^2} \right]^{1/2} \ ,$$

124

$$\Omega = \frac{a}{c(r_+^2 + a^2)} \quad ,$$

$$V = \frac{Qr_+}{c^2(r_+^2 + a^2)} \quad .$$

Equation (4.20) is formally analogous to the well-known thermodynamic relation

$$dM = TdS + \Omega dL + VdQ \tag{4.21}$$

where T is the temperature and S the entropy of a thermodynamic system. With this analogy, K and A should correspond formally to the temperature and the entropy of the black hole respectively. Indeed, the area of the horizon A, as shown in (4.17), possesses the principal property of entropy with the irreducibility of A analogous to the irreducibility of entropy in thermodynamics.

It was later when Bekinstein *et al.* proved that the similarity is not just a formal analogy but an identity in which the surface area A should be identified with the black-hole entropy, or, strictly speaking, where the black-hole entropy can be expressed by

$$S = \frac{kc}{4\hbar G} A \quad .$$

where k is Boltzmann's constant. Thus, we can transform (4.20) into (4.21) provided the black hole temperature is defined as

$$T \equiv \frac{G\hbar}{2\pi kc} K \quad . \tag{4.22}$$

Hence, by identifying eq. (4.22) with eq. (4.21), parameter T defined in (4.22) must be regarded as the thermodynamic temperature.

If this is true, a problem emerges because for any thermodynamic system with temperature T, there must be black-body radiation corresponding to T, whereas in contradiction, the principal property of a black hole is the impossibility of the emission of any matter or radiation from within through the horizon.

125

The contradiction was resolved by Hawking in 1974 when he proved that a black hole should indeed radiate photons and other particles with a spectrum corresponding to the temperature (4.22). The key idea is to take quantum effects into account. According to quantum theory, quantum processes of creation and annihilation of virtual pairs occur constantly in the vacuum state. The existence of these processes called vacuum fluctuations have already been proved by the experiments of electrodynamics, so that it is only natural that there exist virtual particles produced by such fluctuations in the neighbourhood of the horizon outside a black hole. For an observer far away from the black hole, some of the virtual particles have negative energy. Through the tunnel effect, particles with negative energy can penetrate the horizon and enter the black hole, causing the mass of the black hole to decrease. This process is equivalent to emitting a particle with positive energy towards infinity. The opposite or inverse processes causing mass increase also occur, implying that fluctuations alone in a local region cannot change the mass of a black hole. However, if one takes into account the fact that a black hole is formed by the collapse of a star, the formation process of a black hole by gravitational collapse is then not invariant under time inversion. This asymmetry in time direction indicates the existence of emissions from black holes as the net influence of vacuum fluctuations.

In principle, black holes can stably emit any kind of matter provided it can be absorbed by them as well. The emissions constitute a thermal spectrum, namely, the emission spectrum under thermal equilibrium. When temperature is low, only the particles with zero rest mass (such as photons, neutrinos) are emitted by black holes. With increase in temperature the different particles in a sequence of increasing mass are gradually emitted in large amounts. The thermal equilibrium temperature T of black holes is determined by (4.22). The emission rates are not related to the three kinds of interactions (strong, weak and electromagnetic interactions), but depend on gravity (see eq. 4.22). This indicates that gravity dominates over all other interactions in the neighbourhood of a black hole despite the quantum effects which are taken into account.

For Schwarzschild black holes, we can transform the formula (4.22) into following form:

$$T = \frac{\hbar c^3}{8\pi GKM} = 0.62 \times 10^{-7}\left(\frac{M_0}{M}\right) \text{ K} \quad .$$

It follows that a black hole with large mass has a low temperature and vice versa. The temperature of a black hole of about one solar mass is only 10^{-7} K, and in such a case, the emission can be regarded as slow evaporation, that is, the mass of the black hole becomes smaller extremely slowly. For a small black hole with mass of 10^{14} g the temperature is about 10^{12} K, and the emission rate is extremely large. If we consider only light emissions, then according to the formula for black-body radiation, the emission rate of black holes would be $A\sigma T^4$, where σ is the Stefan's constant. Thus, the rate of decrease of the mass of a black hole due to black-body emission is

$$\frac{dM}{dt} = -\frac{\sigma}{c^2} AT^4 = -\frac{\beta}{M^2} \quad ,$$

where formulae (4.22) and others have been used with

$$\beta = \frac{\hbar c^4}{(30.8)^3 \pi G^2} \quad .$$

By integrating the above equation, it is easy to calculate that the lifetime determined by radiation intensity for a black hole of initially one solar mass is

$$t = \frac{1}{3\beta} M_0^3 = 8.4 \times 10^{-24} M_0^3 \text{ s} \quad ,$$

where M_0 has units in grams. It can be seen that black holes of small mass have extremely short lifetimes and release a great deal of energy as radiation over a very short time not unlike a strong explosion.

According to quantum theory, a black hole is neither an absolute stationary state nor even a relative stationary state; it is an excited state of gravity. Emission by black holes is basically analogous to the spontaneous emission by an excited state, because the emissions

127

'triggered' by virtual particles in vacuum in quantum theory are all spontaneous emissions of excited states. After a star achieves the state of a black hole, matter is constantly produced as emissions.

The emissions of a black hole make it possible to regard T defined in (4.22) as the thermodynamical temperature, but the problem does not end here. Since the temperature of a black hole is inversely proportional to its mass M, a black hole should be a system with negative specific heat: as heat flows into it, the temperature of a black hole should decrease instead of increase, and as heat is radiated the temperature increases. The direct consequence of this property is that a system containing black holes cannot reach stable equilibrium. When the temperature of a black hole is the same as that of its environment, the accretion and the emission of the black hole have equal rates, i.e. the black hole does not grow nor diminish. However, as long as the accretion prevails, even if slightly, over the emissions due to fluctuations, a decrease in the temperature of the black hole occurs which in turn diminishes the emissions further. Consequently, the difference between the rates of accretion and emission becomes larger and the temperature of the black hole becomes lower than that of the surroundings, disrupting the equilibrium state. Conversely, if the emission is initially only a bit larger than the accretion, this will lead to the result that the black hole temperature becomes progressively higher than that of the surroundings.

In conclusion, since emissions by black holes have been explained, it is possible for us to regard eq. (4.20) as the basic law of thermodynamics of black holes. On the other hand, the introduction of black holes encounters the more difficult problem on the concept of thermodynamic equilibrium.

These discussions show that ideas concerning black holes are deeply related to gravitation, quantum theory, thermodynamics and statistical physics and are of great significance to these basic concepts. There is no doubt that it is very important to reveal the relations between these basic theories by studying black holes.

4-5   Identification of Black Holes

After pulsars were discovered and identified with neutron stars, the confirmation of the existence of black holes by observation became an urgent matter. We wish now to discuss what properties enable direct comparison with astrophysical observations to be made.

Neutron stars and black holes are two kinds of compact stars which have physical parameters with very close values, such as their radius, angular velocity and magnetic field strength. Data on neutron stars and black holes are shown in Fig. 4.3.

Although there are many similarities, an important difference in mass exists between neutron stars and black holes: the mass of a neutron star is always less than the absolute critical mass $(3.2\,M_\odot)$. The possibility that a star of mass smaller than the critical mass

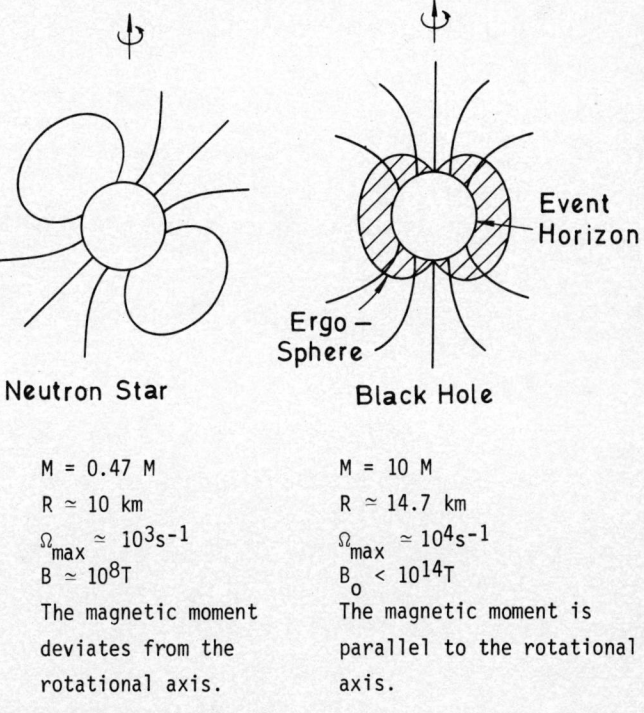

Neutron Star

Black Hole

Event Horizon

Ergo – Sphere

$M = 0.47\,M$	$M = 10\,M$
$R \simeq 10$ km	$R \simeq 14.7$ km
$\Omega_{max} \simeq 10^3 s^{-1}$	$\Omega_{max} \simeq 10^4 s^{-1}$
$B \simeq 10^8 T$	$B_o < 10^{14} T$
The magnetic moment deviates from the rotational axis.	The magnetic moment is parallel to the rotational axis.

Fig. 4.3 Comparison between neutron star and black hole

collapsing to a black hole is extremely small although this cannot be ruled out in principle, because it is necessary for the collapsing of the star to have so much momentum that the potential barrier of the equilibrium state of the neutron star may be overcome. In any case, it may be said that the mass of a star that can collapse to form a black hole, in general, should be greater than 3.2 $M_{\odot}$.

Another basic difference between neutron stars and black holes is in their electromagnetic structure. For neutron stars, the direction of the magnetic moment can deviate from the rotational axis and this inclined magnetic moment was used earlier to explain the periodic emission of pulsars. Stationary black holes, on the other hand, have only axial magnetic moments. Thus, when matter falls into the neighbourhood of a black hole although there will be emissions of short time intervals (the time scale is about that for a revolution of the particles round the black hole), the radiation cannot have periodic structure.

In order to verify the theory by observation, it is therefore necessary to measure mass, angular velocity and electromagnetic properties (i.e. the structure of the magnetosphere) of the compact stars. Studies are also done on the internal structure of neutron stars. It is obvious that pulsars cannot be used for these measurements, since

1. the emission process of pulsars is very stable, and no change can be observed except for a sudden change of only a few pulse periods;

2. radio pulsars are all single stars (except PSR1913+16) so that it is impossible to measure their masses directly.

To find an isolated black hole appears even more hopeless, because no light ray is emitted from a black hole. As for the influence of a black hole on a distant star, it would be notable only when the black hole on and the distant star lie along the same line of sight, which, needless to say, is even more of a rarity.

This situation impels us emphatically to discuss black holes belonging to binary star systems, because in the event that the other member of a binary star system is a normal star, it is possible to have more observable effects.

The gravitational equipotential surfaces of a binary star system consisting of a normal star M_1 and a compact star M_2 are shown in Fig. 4.4, where L_0, L_1, L_2 are called Lagrange points. The equipotential surface in the form of an 8 is called a Roche lobe. It ought to be mentioned that we have assumed the following for the calculation of the equipotential surfaces in Fig. 4.4:

1. the masses of the two stars are concentrated at two points;

2. the orbits of the two stars are circles;

3. rotational axes are perpendicular to the orbital plane;

4. the period of rotation of M_1 is equal to that of the revolution.

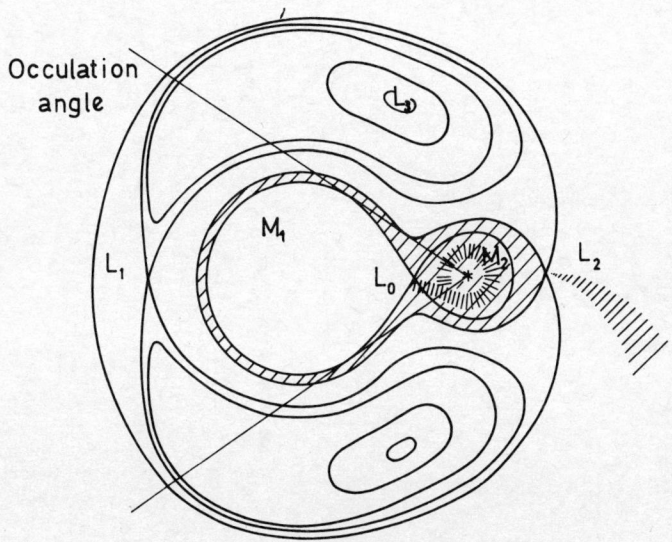

Fig. 4.4 Gravitational equipotential surfaces of a binary star system consisting of a normal star M_1 and a compact star M_2

131

At a certain stage in the evolution of binary stars, the normal star M_1 can fill half of the Roche lobe, and matter from the normal star flows to and is accreted by the compact star through the Lagrange point L_0. When the accreted matter flows through the intense gravitational field, there will be emissions.

Now let us discuss some general features of this binary star system. One feature is the existence of an upper limit on the accreting luminescence, because if the radiation is too intense, the outward light pressure would be so great that the accreting process could be stopped and, thus, the radiation would decrease. Suppose the luminosity of the compact star is L, then the outward light pressure on electrons at r would be $L\sigma_e/4\pi r^2 c$, where σ_e denotes the Thomson cross-section of electrons. If the light pressure equals the gravitational force on a hydrogen atom, we have

$$\frac{L\sigma_e}{4\pi r^2 c} = \frac{Gm_p M}{r^2} \quad,$$

or

$$L = L_E \equiv \frac{4\pi Gm_p M}{\sigma_e} = 1.3 \times 10^{38}\left(\frac{M}{M_0}\right) \text{ erg s}^{-1} \quad,$$

where m_p is the mass of the proton, and the Eddington limit L_E is the upper limit of the accreting luminescence.

If the potential energy released by the accreting matter is equal to L_E, then

$$\frac{GM_2}{r_2}\frac{dM}{dt} = L_E \quad,$$

where dM/dt is the accretion rate, and r_2, the radius of the compact star. Therefore, the accretion rate is

$$\left(\frac{dM}{dt}\right) \leqslant \frac{L_E r_2}{GM_2} \simeq 10^{-8} M_\odot/\text{year} \quad.$$

Since part of the energy should be carried away by neutrinos, and another quite large part would probably not be radiated, the real accretion rate may be higher and reach

$$(\frac{dM}{dt}) \simeq 10^{-6} M_{\odot}/year.$$

Moreover, for a close binary system with filled Roche lobe, there must be occultation as long as the inclination i of the orbit plane is not too small, and the time of occultation has the same magnitude as the period of the binary star. The angles of occultation for different parameters are shown in Table 8. The definition of angle of occultation is

$$\phi = \frac{\text{time of occultation}}{\text{orbit period}} \times 180^{o}$$

Table 8. Occultation angles of close binary systems

$q=M_2/M_1$	$i=90^{o}$	$i=80^{o}$	$i=70^{o}$	$i=60^{o}$	$i=50^{o}$	$i=40^{o}$	$i=30^{o}$	$i=20^{o}$
1.0	22.00	19.81	10.46	-	-			
0.8	23.30	21.29	13.25	-	-			
0.6	25.03	23.22	14.40	-	-	region without		
0.4	27.56	26.00	20.41	-	-	occultation		
0.3	29.42	28.00	23.09	9.58	-			
0.2	32.09	30.85	26.69	17.15	-	-	-	-
0.15	34.00	32.88	29.16	21.16	-	-	-	-
0.1	36.72	35.73	32.52	26.04	10.77	-	-	-
0.05	41.32	40.52	37.97	33.15	24.50	-	-	-
0.02	47.16	46.54	44.60	41.13	35.58	26.70	-	-
0.01	51.29	50.77	49.17	46.37	42.10	35.92	27.01	12.59
0.005	55.13	54.69	53.35	51.04	47.63	42.98	36.97	29.83
0.001	62.87	62.55	61.63	60.06	57.87	55.07	51.83	48.54

It can be seen from the above table that if the distribution of inclinations of the orbit plane is homogeneous, then occultation does not occur for about 4/9th of close binary systems. R. Giacconi and his colleagues, employing the Uhuru satellite for X-ray observation, have found a number of X-ray binary systems since 1971. All these systems have very short periods, generally smaller than five days, and are close binary systems consisting of one normal and one compact star.

These discoveries gave us the first chance to measure the mass of compact stars and, therefore, to test effectively the theory of neutron stars and black holes.

These X-ray binary stars can be clearly divided into two different types: one has X-ray emission with regular pulses and many characteristics similar to that of radio-pulsars, while the other emits X-rays without regular pulses, but with luminosity variation time-scales of several ms. The typical emission forms of these two types of sources are shown in Fig. 4.5.

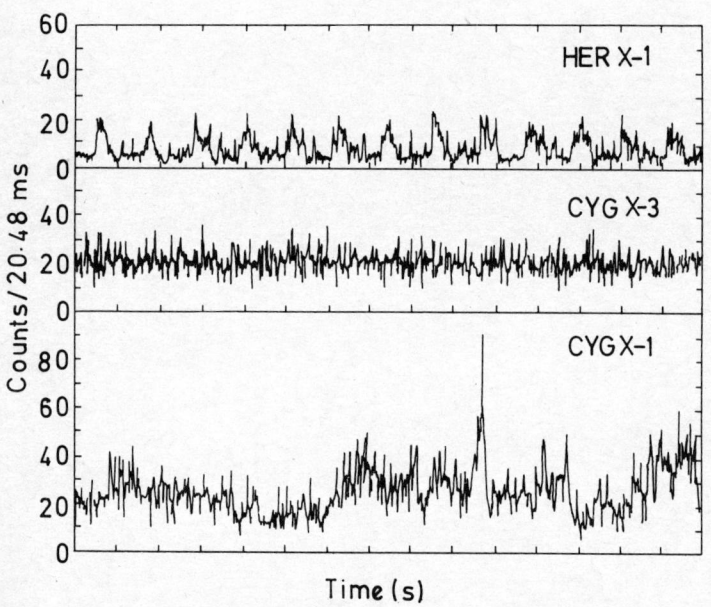

Fig. 4.5 Observational data of the two kinds of X-ray binary stars

We know from the discussion that probably only neutron stars have regular emissions of pulses but black holes have none, because a rotating neutron star has a periodically changing magnetic moment in contrast to a rotational black hole which has no such feature, its magnetic moment being always parallel to its rotational axis. Therefore, we may roughly conclude that X-ray sources with periodical pulse structure should probably be neutron stars, and others with irregular luminosity variations of short time-scales ought to be black holes.

To examine this conclusion one must measure mass. If all masses of the first type of source are smaller than the critical mass and those of the second type larger, the predictions of the theory on black holes is tested at the first step. We are going to discuss the first and second types of X-ray sources in the next two sections 4.6 and 4.7 respectively.

4-6   X-ray Pulsars in Close Binary Star Systems

The two X-ray sources in binary stars, Her X-1 and Cen X-3, have emissions which are typical pulsation structures. The pulsation period of the former is 1.23 s and that of the latter is 4.84 s. They have the characteristics of typical close binary stars:

1. there are typical eclipses in X-ray emission;

2. the pulsation period P_0 is modulated by the Doppler effect due to the orbital motion of the source.

The basic results of observations on Cen X-3 are given in Fig. 4.6 in which figure C apparently indicates that there are eclipses in X-ray emission, the period of the binary star is T = 2.087 days, the eclipses (also called low states) last 0.55 days, and the transition time from high state to low state is about 0.04 days.

The X-ray source moves along a circular orbit. It follows that the source in a low state is farther from us and its pulse would be delayed by 39.7 s, while in a high state, the source is near and the time for receiving the pulses would be earlier by 39.7s. The curve of

135

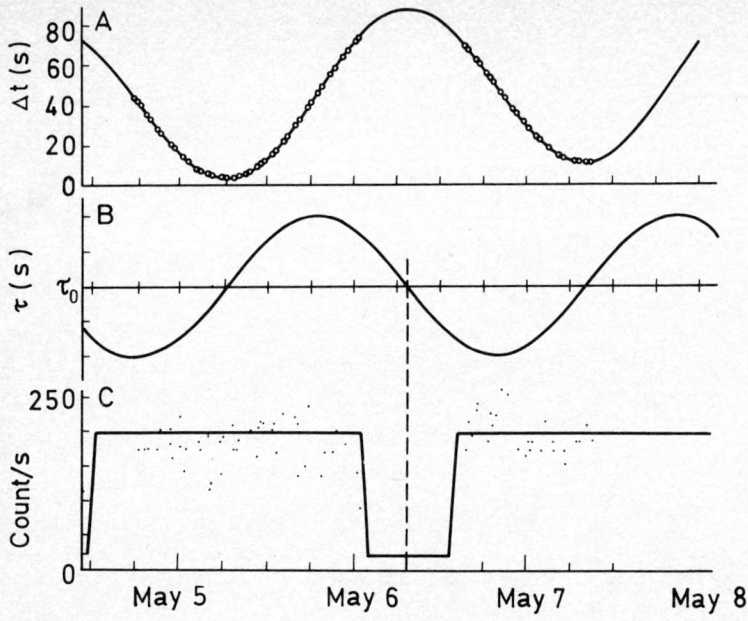

Fig. 4.6 X-ray data of Cen X-3

the arrival time is

$$\Delta t = at + b \sin \frac{2\pi}{T}(t - t_0)$$

where

$a = 0.00198 + 0.000001$
$b = 39.7466 \pm 0.0362\,s$
$T = 2.08707 \pm 0.00025$ days.

Here, $39.7c$ expresses the projection of the orbital radius of the X-ray source along the line of sight.

Figure 4.6B shows the pulsation period modulated by the Doppler effect due to the orbital motion of the source. It is presented in the form of a sine curve, and its amplitude is directly proportional to the projection of the velocity of the X-ray source along the line of sight, which proves that $v_x \sin i = 415.1\ km\,s^{-1}$.

136

That all curves in Fig. 4.6A and B approximate to the sine, indicates that the orbit, with very small eccentricity $e < 0.05$, is quite close to being a circle.

Now, as long as the inclination of the orbital plane is given, we can get, according to the period, the eclipse time and projection of the orbital velocity, the different parameters of the binary star. These parameters for Cen X-3 are as in Table 9.

Table 9. Parameters for Cen X-3

I	$m_x/M_\odot$	$m/M_\odot$	$a/R_\odot$	$R/R_\odot$	v_x	v_m
90°	0.275	16.0	17.4	12.7	415.1	7.14
80°	0.250	16.6	17.7	13.3	421.5	6.32
60°	0.194	24.2	19.9	15.5	479.3	3.83

In Table 9, m_x and m are the masses; and v_x and v_m are the speeds of the X-ray and normal stars along the circular orbit respectively, a is the orbital radius and R the radius of the normal star.

A completely analogous analysis is available for Her X-1, and the results are given in Table 10 below:

Table 10. Parameters of Her X-1

i	$m_x/M_\odot$	$m/M_\odot$	$a/R_\odot$	$R/R_\odot$	v_x	v_m
90°	1.20	2.1	8.9	3.8	169.0	96.3
80°	0.78	1.8	8.2	3.7	171.6	73.8
60°	0.17	1.6	7.2	4.1	195.2	20.5

These results indicate that all masses of these X-ray stars are smaller than the absolute critical mass, which agrees with the theoretical prediction on neutron stars. However, the mass of Her X-1 is obviously greater than the critical mass which Oppenheimer and Volkoff calculated by using the free neutron gas model. Therefore, it can be said that the existence of repulsion between neutrons has been verified in astrophysics. The model of neutron stars also contributes to a better understanding of the following:

1. In the X-ray range, the emission power of sources is $10^{37}\,\mathrm{erg\,s^{-1}}$ and does not exceed the Eddington limit of luminosity; in other words, the luminosity is just determined by the Eddington limit.

2. The X-rays are emitted in the form of pulses, the period of which is about 1 s. It is recognized, as are pulsars, that the periodicity of X-ray emissions is due to the rotation of neutron stars.

3. The pulsation periods of X-rays shorten with time.

The third point differs from radio pulsars, which have periods which increase slowly with time, i.e.

$$\frac{dP}{dt} > 0 \quad .$$

Therefore, it can be considered that the energy sources of pulsars are provided by the rotational energy of neutron stars. However, in X-ray sources the pulsation period decreases slowly, i.e.

$$\frac{dP}{dt} < 0 \quad .$$

Thus, the rotational energy of X-ray sources also increases slowly. The results of observations on the decrease of the pulsation period of Cen X-3 are shown in Fig. 4.7.

The emissions of an X-ray source and the increase of its rotational energy is actually from accretion, or in other words, matter flows continuously into the compact star from the normal star filling the Roche lobe and the emissions of X-rays are translated from the gravitational energy of the accreted matter. The angular momentum of the accreted matter causes the rotational energy of the X-ray star to increase.

138

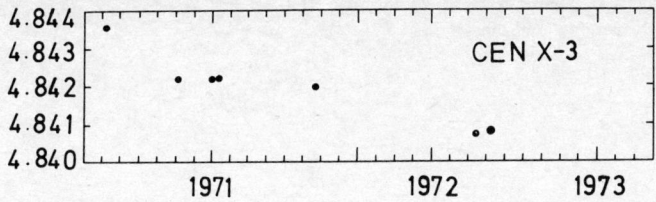

Fig. 4.7 Pulsation periods of Cen X-3 shorten with time

Generally speaking, as accreted matter with certain angular momentum falls into a compact star, a matter distribution, called the accretion disc, in the form of a disc around the compact star is formed. The matter in an accretion disc is in Kepler motion around the compact star. When the orbital radius, decreasing due to the viscosity, equals that of the neutron star, the matter falls into the neutron star. In this case, the angular momentum equation of the neutron star is

$$\frac{dm}{dt} (GMR)^{1/2} = 1 \frac{d\omega}{dt} \quad,$$

where M is the mass of the neutron star, I the moment of inertia and R the radius. The left-hand side of the above formula expresses the angular momentum of matter falling into the neutron star in unit time, while the right-hand side expresses the increase rate of the angular momentum of the neutron star. From observation data on $d\omega/dt$, we have

$$\frac{dm}{dt} \lesssim 1.5 \times 10^{-10} M_{\odot}/year \quad.$$

On the other hand, it can be deduced from the energy relations that if the luminosity of $10^{37} ergs^{-1}$ of X-rays is completely provided by the gravitational energy of the accreted matter, the absorption rate would be at least

$$\frac{dm}{dt} \simeq 1.5 \times 10^{-9} M_{\odot}/year \quad.$$

The contradiction between the above two formulae indicates that the developed model of an accretion disc is too much of an approximation. The influence of magnetic fields should be taken into account. Due to

139

this influence, the motion of the accreted matter is not a complete Kepler motion, but is along magnetic lines falling into a neutron star. Thus, the influence removes the limit on the equation for angular momentum. The existence of strong magnetic fields in Her X-1 has already been verified by observations. For example, the line spectrum of 58 keV found in X-ray emissions of Her X-1 can be explained by emissions of the Landau energy level in the magnetic fields.

Recent observations of the line spectrum of 110 keV has further verified this point. Since the difference with respect to the Landau energy level is $\hbar eB/mec$, it is easy to evaluate the magnetic field strength in Her X-1 to be about 10^8 T.

On the other hand, the revolution period of Cen X-3 decreases with time, and in 1971 the rate of change was $\Delta T/T \simeq 3.5 \times 10^{-5}$. If the rate is caused by mass loss, the loss rate of mass would be

$$\frac{dM}{dt} \gtrsim 10^{-3} M_{\odot}/\text{year} \quad .$$

It follows that only a few parts of the lost mass of the normal star is absorbed by the compact star.

4-7 Close X-ray Binary Star Systems with Fluctuating Luminosities

This type of X-ray sources can also be one of the two components of a binary system. The X-ray luminosity is also about 10^{38} ergs^{-1}. Nevertheless, the sources do not emit regular pulses, and their X-brightness fluctuates (short time-scale variability). For example, the X-brightness could vary several times like X-ray flares.

According to the analysis of the last section, it is still possible that the X-rays of such sources originate from the accretion of a compact star. That the brightness fluctuates on a very short time-scale indicates that the region of X-ray emission should be extremely small.

Cyg X-1 is a typical fluctuating X-ray source, whose luminosity is $L \sim 10^{37}$ erg s^{-1}. In the early observations on it, the characteristics of binary stars were not found. For example, no eclipses

were observed. Since it also does not have periodic pulses, the
Doppler effect cannot be used for identification if it is a binary star.
Therefore, if Cyg X-1 is a binary star, its orbital plane inclination
i should be so small that the eclipses cannot be observed. There are
other facts, indeed, to support this conclusion about inclination i.
One of them is the existence of a cut at the low frequency end of the
spectrum. It can be seen in Fig. 4.8 that the low frequency absorption
of Cyg X-1, compared with Her X-1 and Cen X-3, is lower since the
X-rays have only to pass through a thinner absorption layer due to the
smaller i.

The final observational evidence for Cyg X-1 being a binary star
is that HDE226868 and Cyg X-1 have been identified as belonging to

$$\frac{d\epsilon}{dE} = CE^{-\alpha} \exp\left\{ -\left(\frac{E_A}{E}\right)^{\frac{8}{3}} \right\}$$

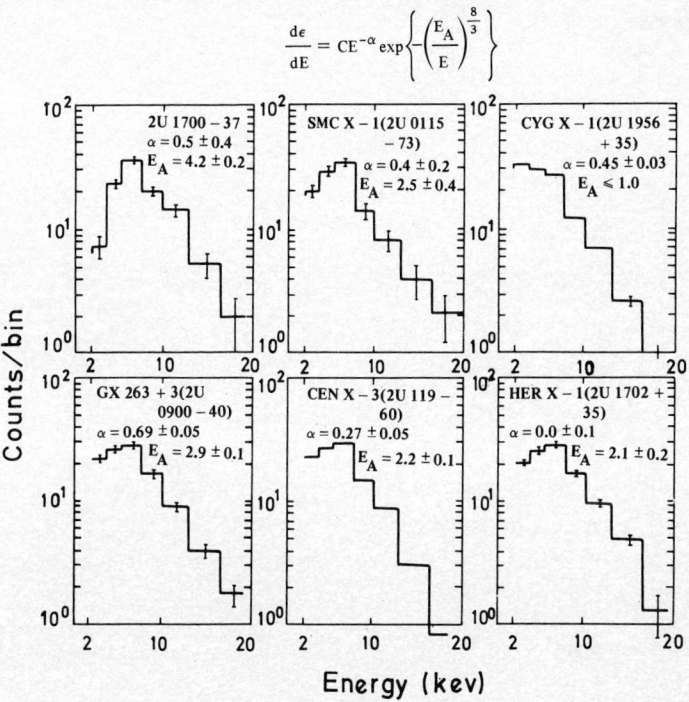

Fig. 4.8 Spectrum of six X-ray binary systems

141

the same binary system. HDE226868 is a Bo supergiant, whose spectral lines show the typical Doppler effect of binary star. Fig. 4.9
shows its velocity-phase curve.

It has been proved by observations that there are gas streams in
the region between HDE226868 and Cyg X-1. This directly indicates
that Cyg X-1 accretes matter from HDE226868.

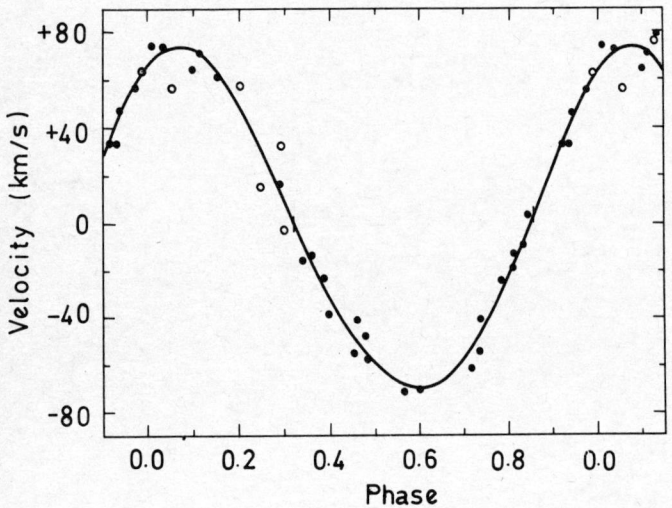

Fig. 4.9 Velocity-phase curve of HDE226868

From the eclipses, the Doppler effect and others, the inclination
i of Cyg X-1 has been estimated to be about 26°. According to these
data, we get the following Table 11 on the estimates of the mass of
Cyg X-1:

142

Table 11. Parameters of Cyg X-1

Cyg X-1	HDE226868	Authors
$M_2/M_\odot > 5.5$	$M_1/M_\odot \simeq 20$	Brucato-Kristian
$10.0 \leqslant M_2/M_\odot \leqslant 18$	$16 \leqslant M_1/M_\odot \leqslant 23$	Hutchings *et al.*
$7.8 \leqslant M_2/M_\odot \leqslant 17$	$10 \leqslant M_1/M_\odot \leqslant 22$	Sunyaev *et al.*
$10.0 \leqslant M_2/M_\odot \leqslant 20$	$25 \leqslant M_1/M_\odot \leqslant 35$	Bolton

All the results obtained in different ways, as we can see, have proved that the mass of Cyg X-1 is greater than the absolute critical mass of neutron stars.

To sum up, Cyg X-1 shows three key characteristics: the luminosity $L \gtrsim 10^{37}$ erg s^{-1}, the X-brightness fluctuates with a very short time-scale of the order of several ms, and the mass is greater than $3.2\ M_\odot$, agreeing with previous predictions on black holes.

REFERENCES

Monograph or collected works on Black Holes

B. De Witt, C. De Witt, ed., Black Holes (Gordon and Breach, 1973).

M. Rees, R. Ruffini, J.A. Wheeler, Black Holes, Gravitational Waves and Cosmology (Gordon and Breach, 1974).

H. Gursky, R. Ruffini, ed., Neutron Stars, Black Holes and Binary X-ray Sources (Reidel, 1975).

R. Giacconi, R. Ruffini, ed., Physics and Astrophysics of Neutron Stars and Black Holes (North-Holland, 1978).

Monograph or collected works on Critical Mass

C.E. Rhoades, R. Ruffini, Phys.Rev.Lett., 32, 324 (1974).

G.R. Blumenthal, W.H. Tuckev, Annual Review of Astronomy and Astro-physics, 12, 150 (1974).

Chapter 5

GRAVITATIONAL WAVES

5-1   Gravitational Fields of 1/r Type

It has been pointed out in section 1.5 that what is called an inertial force is caused, according to Mach's Principle, by the direct interaction between matter on the universal scale and a body moving in a local region, and not by absolute space. We now analyse what properties the direct interactions mentioned above should have, if any at all.

Consider an external force $\vec{F}$ acting on a body. According to Newtonian mechanics, the force causes the body to move with acceleration $\vec{a} = \vec{F}/m$ relative to an inertial frame K. In a non-inertial frame K' moving with the body, the body's acceleration $\vec{a}' = 0$. It is obvious that there should be another force $\vec{F}'$ besides $\vec{F}$, acting on the body, and $\vec{F}' = -\vec{F}$, so that

$$m\,\vec{a} = \text{total force} = \vec{F}' + \vec{F} = 0 \ .$$

$\vec{F}'$ is a fictitious force in the Newtonian system, but from Mach's Principle, $\vec{F}'$ originates from the acceleration of matter in the universe. The main difference between the inertial frame K and the non-inertial frame K' is that universal matter has no accelerating motion relative to K, but on the average has an acceleration

144

relative to the frame K'.

Now let us discuss, from a simple viewpoint, how the force $\vec{F}'$
comes about. We start by analysing the contribution of a body of mass
M to the force $\vec{F}'$, and then sum up. When M moves with an acceleration $\vec{a}$ relative to m, what effect would M have on m? Firstly, there is a general static attraction GMm/r^2, where r is the
distance between M and m. The inverse square law indicates that the
effects of near bodies are more important. Indeed, the attraction of
the Earth on bodies near the ground, for example, outweighs the effects
of all other bodies. The sum of the static forces from other parts of
the universe is actually zero, since the distribution of mass is isotropic, and their effects cancel out. Secondly, the force determined
by the inverse square law is not concerned with the acceleration and,
therefore, would not contribute to $\vec{F}'$. This inverse square force is
similar to the Coulomb interaction between two charges. However, there
is another electromagnetic force between two charges which move relative to each other with an acceleration $\vec{a}$. The magnitude of the
force caused by electromagnetic induction is

$$\vec{F} = \frac{q_1 q_2 \vec{a}}{c^2 r} \tag{5.1}$$

Similarly, we can infer that there may be this kind of induced
force, the magnitude of which is

$$\vec{F} = \frac{GMm\vec{a}}{c^2 r} \ , \tag{5.2}$$

between two bodies moving with relative acceleration $\vec{a}$.

The existence of forces of type (5.2) coincides with Mach's point
of view. Firstly, as $\vec{F}'$ concerns acceleration, it exists only for
the frame K'; there is no $\vec{F}'$ in the frame K. Secondly, $\vec{F}'$ is of
the 1/r type, that is, the induced force caused by distant bodies is
much larger than their static force. The above two properties are
characteristics of radiation fields. According to electrodynamics, the
necessary condition for causing radiation is the presence of an

acceleration of a charged body; the radiation field in the region as $r \to \infty$, produced by a system near $r = 0$, falls by $1/r$. Therefore, one concludes that if Mach's theory is correct, there should be radiated gravitational fields, namely, gravitational waves. Accelerated matter can radiate gravitational waves in the same way that accelerated charge radiates electromagnetic waves.

General relativity agrees with Mach's Principle on the point that universal matter influences local motion, but differs because general relativity holds that the influences of universal matter on local motion are not direct interactions, but indirect effects, by determining which frame among local frames is the inertial one. This is despite the fact that the existence of gravitational waves inferred from Mach's Principle coincides with the general result from general relativity.

We can still use Mach's Principle for semi-quantitative estimates. Imagine a binary system of one solar mass consisting of two neutron stars with a distance $R \sim 10^6$ km between them. Their acceleration is

$$a = \frac{GM_\odot}{R^2} \approx 10^4 \text{ cm s}^{-2} \quad .$$

If the binary system is 100 light years away, namely $R \sim 10^{20}$ cm, then, according to formula (5.2), the system would produce an acceleration on the Earth:

$$A \approx \frac{GM_\odot a}{c^2 r} \approx 10^{-11} \text{ cm s}^{-2} \quad .$$

This value is too small to measure directly at the existing technical level. We cannot measure an acceleration that is added to that of all bodies in the local region of the Earth but only the tidal effect caused by gravitational waves. Since the order of the period of the gravitational waves is

$$T \sim R / \frac{GM_\odot}{R} \sim 10^4 \text{ s} \quad ,$$

the wavelength would be

$cT \sim 10^{14}$ cm .

Hence, the tidal force acting on a body of unit length is about

$$\frac{A}{cT} \quad 10^{-24} \ s^{-2} \quad .$$

Thus, the distance moved between two ends of the unit length is about

$$\frac{1}{2} \ (\frac{A}{cT})(\frac{T}{2})^2 \sim 10^{-17} \quad .$$

In other words, to measure the gravitational waves radiated by the two stars, the relative accuracy of measuring distance must exceed 10^{-17}. This is impossible at the moment.

5-2 Deviation Equation

It has been pointed out in the preceding section that we can only use the relative motion between two neighbouring bodies, i.e. the effects caused by the tidal force, to detect gravitational waves. For further discussion of gravitational waves, we now consider the deviation equation — the fundamental equation for describing the tidal force.

Let us discuss a very simple case first. For two particles with position coordinates x_1 and x_2 respectively, in a weak gravitational field described by Newtonian gravitation potential $\phi(x)$, the equations of motion for the two particles are

$$\frac{d^2 x_1^i}{dt^2} = - \frac{\partial}{\partial x^i} \phi(x_1) \quad ,$$

$$\frac{d^2 x_2^i}{dt^2} = - \frac{\partial}{\partial x^i} \phi(x_2) \quad .$$

The relative acceleration between the two particles is

$$\frac{d^2 x_1^i}{dt^2} - \frac{d^2 x_2^i}{dt^2} = - \frac{\partial}{\partial x^i} \phi(x_1) + \frac{\partial}{\partial x^i} \phi(x_2) \quad .$$

If they are quite near, then $x_2^i = x_1^i + \delta x^i$. It is easy to find an equation for δx^i from the above equation as follows:

$$-\frac{d^2 \delta x^i}{dt^2} = -\frac{\partial}{\partial x^i}\phi(x_1) + \frac{\partial}{\partial x^i}\phi(x_1^i + \delta x^i) = +\frac{\partial^2 \phi}{\partial x^i \partial x^j}\delta x^j$$

This is the fundamental equation, i.e. deviation equation, of the relative motion between x_1 and x_2 caused by the tidal force.

By using the Riemann tensor, the deviation equation can be written as

$$\frac{d^2 \delta x^i}{dt^2} = R^i_{0j0}\,\delta x^j \quad . \tag{5.3}$$

From this formula, we can measure the Riemann curvature at a point from the deviation properties of its neighbour points. This essential characteristic of spacetime curvature is independent of the coordinate system.

5-3 Essential Properties of Gravitational Waves

As early as 1918, Einstein predicted the existence of gravitational waves by general relativity.

A wave in physics, generally speaking, is a propagating process of some perturbation. Gravitation is a physical object described by the spacetime metrics $g_{\mu\nu}$, so that a gravitational wave can be regarded as a propagation of some perturbation on the metrics $g_{\mu\nu}$. For example, a spacetime region without gravitational fields is flat and is described by Minkowski's metric:

$$\eta_{\mu\nu} = \begin{pmatrix} -1 & 0 & 0 & 0 \\ 0 & 1 & 0 & 0 \\ 0 & 0 & 1 & 0 \\ 0 & 0 & 0 & 1 \end{pmatrix}$$

If there is a weak gravitational field in the region, the metric can be written as

$$g_{\mu\nu} = \eta_{\mu\nu} + h_{\mu\nu}$$

where the quantity $h_{\mu\nu}$ is very small. Substituting this metric into Einstein's equation of gravitational fields and retaining in the field equation only the terms which are linear in $h_{\mu\nu}$ and their derivatives, omitting all terms of higher order, we obtain the equation concerning $h_{\mu\nu}$:

$$\Box h_{\mu\nu} = 0 \ . \tag{5.4}$$

When

$$t = x^0, \ \ x = x^1, \ \ y = x^2, \ \ z = x^3 \ ,$$

$$\Box = \frac{\partial^2}{\partial x^2} + \frac{\partial^2}{\partial y^2} + \frac{\partial^2}{\partial z^2} - \frac{1}{c^2}\frac{\partial^2}{\partial t^2} \ .$$

Equation (5.4) is a typical wave equation, but the propagating speed of the wave is the speed of light.

Equation (5.4) is very similar to the wave equation

$$\Box A_\mu = 0 \tag{5.5}$$

in electrodynamics, and this is not merely an analogy between (5.4) and (5.5). We know that (5.5) is valid under the gauge condition

$$\frac{\partial}{\partial x^\mu} A_\mu = 0 \ . \tag{5.6}$$

By analogy, equation (5.4) is valid only under the following coordinate condition:

$$\frac{\partial}{\partial x^\mu} h^\mu_{\ \nu} = \frac{1}{2}\frac{\partial}{\partial x^\nu} h^\mu_{\ \mu} \ . \tag{5.7}$$

Thus, coordinate condition (5.7) is to the gravitational field what the gauge condition (5.6) is to the electromagnetic field. There is no doubt that neither the gauge condition (5.6) nor the coordinate condition (5.7) can influence the physical results.

Generally speaking, for different coordinate systems, the difference is not only in $h_{\mu\nu}$, but also in the form of the equation in $h_{\mu\nu}$. Therefore, the many properties of the solution obtained should be dependent on the choice of the coordinate system. Just because of this point, there have been doubts as to whether the wave solution of eq.

149

(5.4) has any significant physical effects and is not merely a formal result. In 1956, Pirani gave a definition of gravitational waves independent of coordinate systems. In the following year, Bondi further proved the existence of plane gravitational wave solutions independent of coordinate systems. In 1959, Bondi, Pirani and Robinson proved that a body at rest can move by the action of pulses of gravitational waves, which means that gravitational waves carry energy and can be detected. It was then, that the existence of gravitational waves became fully affirmed theoretically.

These theoretical analyses indicate that $h_{\mu\nu}$ contain both the formal parts dependent on the choice of coordinate systems and the relevant content on gravitational waves. We are certainly interested in the latter, but we wish to distinguish between the two. We have returned to our old point of view: gravitational forces may be eliminated from a local region by suitable choice of a coordinate system, but tidal forces cannot. Therefore, in order to perceive clearly the relevant content in $h_{\mu\nu}$, let us analyse its tidal force.

Now we will discuss the simplest solution, i.e. the plane wave solution, of eq. (5.4). If the plane waves propagate along the z axis, we have a solution, from eq. (5.4) and condition (5.7) in the following form:

$$h_{xx} = -h_{yy} = \mathrm{Re}\left\{A_+ e^{-i\omega(t - z/c)}\right\} \ ,$$

$$h_{xy} = h_{yx} = \mathrm{Re}\left\{A_x e^{-i\omega(t - z/c)}\right\} \ , \tag{5.8}$$

with other components of $h_{\mu\nu} = 0$, where A_+ and A_x represent two independent modes of gravitational oscillation where the signs $+$ and x denote two different polarizations. What tidal forces would these plane waves produce?

Consider two quite closely neighbouring test particles A and B. In the proper reference frame of this system, A and B are in the plane vertical to the z axis. From (5.3), the separation δx^i between the two particles satisfies the following equation:

$$\frac{d^2 \delta x^i}{dt^2} = R^i_{0j0} \, \delta x_j \quad .$$

Substituting (5.8) into the above formula, we can get

$$\frac{d^2 \delta x^j}{dt^2} = \frac{1}{2} \frac{\partial^2 h^j_k}{\partial t^2} \, \delta x^k \quad . \tag{5.9}$$

If A is taken as the origin, the δx^i are just the coordinates x_B, y_B, z_B of the point B. If $A_x = 0$ in (5.8), then, from (5.9) the following equation can be obtained:

$$\frac{d^2 x_B}{dt^2} = \frac{1}{2} \frac{\partial^2 h_{xx}}{\partial t^2} \, x_B = -\frac{1}{2} \, \omega^2 \text{Re} \left\{ A_+ e^{-i\omega(t - z/c)} \right\} x_B \quad ,$$

$$\frac{d^2 y_B}{dt^2} = \frac{1}{2} \frac{\partial^2 h_{yy}}{\partial t^2} \, y_B = \frac{1}{2} \, \omega^2 \text{Re} \left\{ A_+ e^{-i\omega(t - z/c)} \right\} y_B \quad ,$$

the solution of which is

$$x_B = x_B(0) + \text{Re} \left\{ \frac{1}{2} A_+ e^{-i\omega(t - z/c)} x_B(0) \right\} \quad ,$$

$$y_B = y_B(0) + \text{Re} \left\{ \frac{1}{2} A_+ e^{-i\omega(t - z/c)} y_B(0) \right\} \quad .$$

By analog, when $A_+ = 0$ and $A_x \neq 0$, the solution is

$$x_B = x_B(0) + \text{Re} \left\{ \frac{1}{2} A_x e^{-i\omega(t - z/c)} y_B(0) \right\} \quad ,$$

$$y_B = y_B(0) + \text{Re} \left\{ \frac{1}{2} A_x e^{-i\omega(t - z/c)} y_B(0) \right\} \quad .$$

The modes of relative motion of the test particles caused by the plane waves with two polarizations are shown in Fig. 5.1. These relative motions are the real physical effects caused by gravitational waves. Various detectors of gravitational waves are designed to detect such effects.

5-4 The Radiation of Gravitational Waves

An accelerated charge system emits electromagnetic waves. By

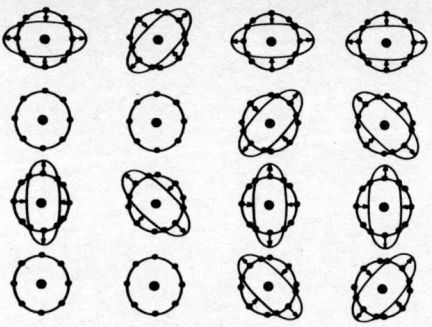

Fig. 5.1   The modes of relative motion of test particles
caused by plane waves with two polarizations.
A is in the centre, B is the peripheral
particle.

analogy, an accelerated matter system should emit gravitational waves.
In the electromagnetic case, the strongest radiation for an isolated
charge system is that of the electric dipole. The definition of an
electric dipole is

$$\vec{d}_e = \sum_i e_i \vec{r}_i \quad ,$$

where e_i and $\vec{r}_i$ are respectively the charge and position vectors of
the ith particle. The intensity of dipole radiation is directly
proportional to $\ddot{\vec{d}}_e$. We can, by direct analogy, define a mass dipole
by replacing e_i in the above formula with mass m_i , that is:

$$\vec{d}_m = m_i \vec{r}_i \quad .$$

The first derivative of $\vec{d}_m$ with respect to time t is the total
momentum of the system

$$\dot{\vec{d}}_m = m_i \dot{\vec{r}}_i = \vec{p} \quad .$$

Due to momentum conservation in an isolated system, $\dot{\vec{d}}_m = \vec{p} = 0$.

152

Therefore, there is no dipole radiation in gravitational physics.

In electromagnetic theory, the radiation second to that of the dipole is that of the magnetic dipole or electric quadrupole. The radiation of a magnetic dipole is determined by its second derivative with respect to time. The magnetic dipole $\vec{\mu}$ of the system is given by

$$\vec{\mu}_{em} = \sum_i \vec{r}_i \times (e_i \vec{v}_i) \quad .$$

Again substituting by analogy, $e_i \to m_i$, the corresponding quantity in gravitation can be obtained as follows:

$$\vec{\mu}_g = \sum \vec{r}_i \times (m_i \vec{v}_i)$$

which is the total angular momentum of the system. From the conservation of angular momentum, there is also no gravitational radiation analogous to the radiation of the magnetic dipole.

The gravitational radiation analogous to the radiation of the electric quadrupole exists and, for an isolated system, is mainly determined by the rate of change of its mass quadrupole. If the mass quadrupole of the system is ·

$$D_{\alpha\beta} = \int \rho(3x^\alpha x^\beta - \delta_\alpha^\beta x^\gamma x^\gamma)dv \quad ,$$

then the rate of energy of the radiation emitted per unit steradian in direction $\vec{n}$ (unit vector) is

$$\frac{d^2E}{dtd\Omega} = \frac{G}{36\pi c^5} \left[\frac{1}{4}(\dddot{D}_{\alpha\beta}n^\alpha n^\beta)^2 + \frac{1}{2}(\dddot{D}_{\alpha\beta}\dddot{D}^{\alpha\beta}) - (\dddot{D}_{\alpha\beta}\dddot{D}_{\alpha\gamma}n^\beta n^\gamma) \right] \quad ,$$

and the total power of the radiation is

$$\frac{dE}{dt} = -\frac{G}{45c^5}(\dddot{D}_{\alpha\beta})^2 \quad . \tag{5.10}$$

5-5 Gravitational Radiation of Binary Stars

A binary star system is the most common celestial system radiating gravitational waves. Since the gravitational field in a binary system is weak, the bodies still have elliptic orbits given by

153

Newtonian mechanics, that is (2.12):

$$r = \frac{a(1 - e^2)}{1 + e \cos\phi}$$

where $a = r_{min}/(1 - e)$ is the major axis of the ellipse. Using (5.10), we can evaluate the total power of radiation of the binary system as follows

$$L \equiv - \frac{dE}{dt} = \left[\frac{32}{5} \frac{G^4}{c^5} m_1^2 m_2^2 (m_1 + m_2)/a^5 \right] f(e) \quad,$$

where m_1 and m_2 are the masses of two stars respectively, and

$$f(e) = \frac{1 + \frac{73}{24} e^2 + \frac{37}{96} e^4}{(1 - e^2)^{7/2}} \quad.$$

The function f is shown in Fig. 5.2. It can be seen from this figure that the radiation power, for a given a, increases with the increasing eccentricity. The reason is that the larger the eccentricity, the larger is the acceleration of the stars so that radiation increases.

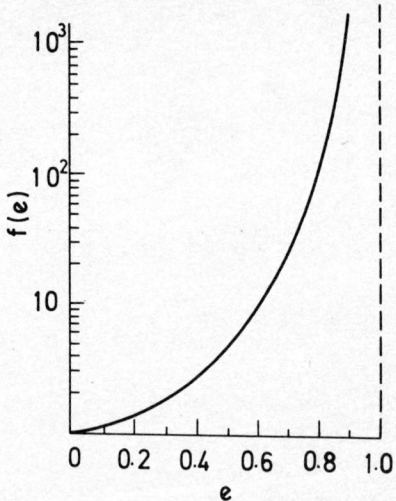

Fig. 5.2 f as a function of e

154

Since the radiation is caused by quadrupole moments, the frequency of gravitational radiation, for a circular orbit, is Ω/π, where Ω is the circular frequency of the rotation of the binary system and is equal to

$$\Omega = \sqrt{\frac{G(m_1 + m_2)}{a^3}} \equiv \frac{2\pi}{T} \quad ,$$

where T is the period of the binary.

There are various harmonic frequencies for an elliptical orbit. By employing Fourier-analysis the total power of the radiation of the frequency $n\Omega/\pi$ is

$$L(n) = \frac{32}{5} \left(\frac{G^4}{c^5}\right) m_1^2 m_2^2 (m_1 + m_2) a^{-5} g(n,e) \quad ,$$

where

$$\begin{aligned}
g(n,e) = \left(\frac{n^4}{32}\right) \Big\{ &[J_{n-2}(ne) - 2e J_{n-1}(ne) \\
&+ \frac{2}{n} J_n(ne) + 2e J_{n+1}(ne) - J_{n+2}(ne)]^2 \\
&+ (1 - e^2)[J_{n-2}(ne) - 2J_n(ne) + J_{n+2}(ne)]^2 \\
&+ \left(\frac{4}{3n^2}\right) J_n^2(ne) \Big\} \quad ,
\end{aligned}$$

where J_n is an nth order Bessel function. The function $g(n,e)$ varying with n is given in Fig. 5.3.

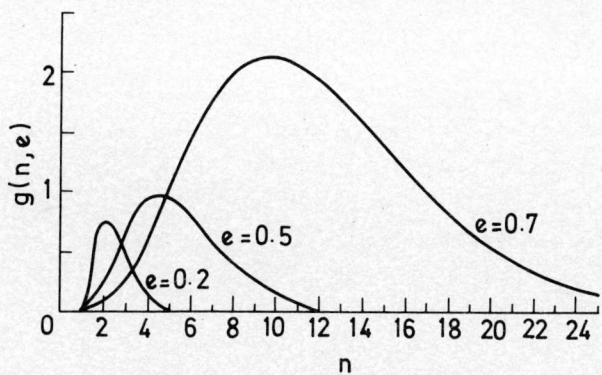

Fig. 5.3 g as a function of n

155

It follows that the larger the eccentricity, the higher will be the maximum frequency of the gravitational radiation, since the acceleration near the perihelion is larger for the larger eccentricity and as the radiation concentrates on the part of the orbit near the perihelion, the time scale in this part is smaller than the period of the whole rotation.

The parameters of the gravitational radiation of some binary systems is given in Table 12.

Even today, the gravitational radiation of binary systems still has not been detected in the laboratory. However, we can demonstrate indirectly the existence of gravitational radiation from its influence on the binary system itself. Because a binary system radiates gravitational waves, its energy should decrease and its parameters should change as well. The total energy of the binary system is

$$E = -\frac{1}{2} Gm_1m_2/a \quad .$$

The loss in energy would gradually cause a to decrease, that is,

$$\frac{da}{dt} = \frac{2a^2}{Gm_1m_2} \frac{dE}{dt} \quad .$$

From Kepler's third law, we can find the rate of change of the period as follows:

$$\frac{1}{T}\frac{dT}{dt} = \frac{3}{2}\frac{1}{a}\frac{da}{dt} = \frac{1}{a}\frac{3a^2}{Gm_1m_2}\frac{dE}{dt}$$

$$= -\frac{96}{5}\frac{G^3}{c^5}m_1m_2(m_1+m_2)\frac{1}{a^4}f \quad .$$

This energy attenuation of binary stars which is caused by radiating gravitational waves is called the gravitational radiation damping.

5-6   Observational Verification of Gravitational Radiation Damping

It is possible, in principle, to test the theory of gravitational radiation damping by observing the change of the orbital period of a binary system, but all binary systems are unsuitable for this test because of non-relativistic factors like the tidal action between two

Table 12. Gravitational radiation of binary systems

Name	Period	m_1/M	m_2/M	Distance (pc)	L (erg s^{-1})	Energy flow on ground (erg s^{-1} cm^{-2})
β Cas	480.00 years	0.94	0.58	5.9	5.6×10^{10}	1.4×10^{-29}
ξ Boo	149.95 "	0.85	0.75	6.7	3.6×10^{12}	6.7×10^{-28}
Sirius	49.94 "	2.28	0.98	2.6	1.1×10^{15}	1.3×10^{-24}
Fu 46	13.12 "	0.31	0.25	6.5	3.6×10^{14}	7.1×10^{-26}
β Lyr	12.925 day	19.48	9.74	330.0	0.057×10^{30}	0.004×10^{-11}
UWCMa	4.295 "	40.00	31.00	1470.0	49×10^{30}	0.019×10^{-11}
β Per	2.867 "	4.70	0.94	30.0	0.014×10^{30}	0.013×10^{-11}
WUMa	0.330 "	0.76	0.57	110.0	0.47×10^{30}	0.032×10^{-11}
UVLeo	0.600 "	1.36	1.25	68.0	0.63×10^{30}	0.012×10^{-11}
V Pup	1.450 "	16.60	9.80	390.0	65×10^{30}	0.36×10^{-11}
i Boo	0.268 "	13.50	0.68	12.0	3.2×10^{30}	18×10^{-11}
YY Fri	0.321 "	0.76	0.50	42.0	0.42×10^{30}	0.20×10^{-11}
SW Lac	0.321 "	0.97	0.83	75.0	1.5×10^{30}	0.21×10^{-11}
WZ Sge	81.00 min.	0.60	0.03	100.0	0.5×10^{30}	0.04×10^{-11}

stars and radiation or stellar wind which can also cause changes in the period.

It is obvious that a binary system which is a good 'lab' for testing the theory of gravitational radiation ought to satisfy the following condition: the relativistic term causing the slowdown of the period >> non-relativistic term which changes the period.

Let us analyse firstly the influence of tides. For a binary system consisting of two stars with radii R_1 and R_2, the rate of period change caused by tides is

$$\frac{1}{T}\frac{dT}{dt} \qquad \frac{3G^3(m_1 + m_2)J^1}{4a^5(1 - e^2)\omega}$$

where mJ^1 is the quadrupole moment caused by the tides, and

$$J^1 \sim R_1^2\left(\frac{R_1}{a}\right)^3 \quad , \quad J^2 \sim R_2^2\left(\frac{R_2}{a}\right)^3 \quad .$$

Therefore, if gravitational radiation damping is larger than the rate of change of the period caused by tides, the following condition must be satisfied:

$$\left(\frac{Gm_1}{c^2 R_1}\right)\left(\frac{a}{R_2}\right) \gg 1 \quad \text{and} \quad \left(\frac{GM_2}{c^2 R_2}\right)\left(\frac{a}{R_2}\right) \gg 1 \quad .$$

Generally speaking, for main sequence stars, $GM/c^2 R \sim 10^{-6}$, and for the compact stars, $GM/c^2 R \sim 10^{-1}$, so that it is easier for compact stars to satisfy the above condition. As long as a is sufficiently large, the condition mentioned can of course be satisfied principally. However, the larger a is, the longer is the period. Consequently, it is impossible to complete the observation of several entire periods and measure the rate of change of the period precisely. In addition, the larger a is, the smaller is the gravitational radiation power, necessitating more accurate time-keeping for the measurement of the rate of change of period of the binary system and simultaneously more difficulties in technique. In brief, binary systems

with smaller a are more suitable for practical measurements and one
can conclude that, in practice, it is only possible for a binary system
consisting of two compact stars to be used to test gravitational radiation damping.

We knew of the existence of compact stars after the discovery of
the pulsar in 1967. However, for some time afterwards there was still
no known radio pulsar consisting of two compact stars as none of about
400 radio pulsars discovered had the obvious characteristics of a binary star system. The majority of known X-ray sources of binary systems consist of one compact star and one ordinary star. These binary
systems are at the stage of strong mass exchange, and cannot be
employed to test the theory on gravitational radiation damping.

At the end of 1974, Hulse and Taylor discovered the radio pulsar
PSR1913+16, which is different from all other known radio pulsars. It
was directly proved by observation that this is a member of a binary
system. The main observational data are given in Table 13 (recorded
in 1978).

Table 13. Parameters of PSR1913+16

R.A. (1950.0)	$\alpha = 19^h 13^m 12^s.474\ 0^s.004$
Dec. (1950.0)	$\delta = 16^0 01'\ 08".02\ 0".06$
Pulse period	$P = 0.05902999526\ (9 \pm 2)$ s
Rate of change of period	$P = (8.64 \pm 0.02) \times 10^{-18}$
Projection of major axis	$a \sin i = 2.3424 \pm 0.0007$ light s
Eccentricity	$e = 0.617155 \pm 0.000007$
Orbit period	$T = 27906.98172 \pm 0.0005$ s

The pulse period of the system is very short, only larger than that of the pulsar NP0532 in the Crab nebula, and occupies the second position in this respect among all pulsars. In addition, the star has a very small orbit period, less than 8 hours, and a very large eccentricity. It is very rare that all these characteristics are found in one system.

It can be inferred from these data that this binary system probably consists of two compact stars, which means that not only is the PSR1913+16 itself a compact star, but so is its companion. The inference is based on the following arguments:

1. this is a binary system and both members do not violate the Roche condition to disintegrate into a disc;

2. no X-rays have been observed, so the companion does not fill the Roche lobe. Besides, stellar wind is also not strong;

3. it can be found from explosion dynamics that the semi-major axis a of the final state is generally larger than that of the initial state.

It is impossible that the preceding stage of the binary star is a system consisting of two stars in the main sequence, otherwise the above three features cannot be satisfied simultaneously.

We know that a system consisting of two stars I and II which are in the main sequence has roughly several evolution stages as shown in Fig. 5.4. The main sequence stage is the initial state (a). The star I with more mass reaches its later phase first and begins to expand, after filling the Roche lobe. The first period of mass exchange begins in (b) and eventually ends in (c). Next, star I expands in a supernova explosion and collapses into a compact star (d). Star II also begins to fill the Roche lobe, forming the second period of mass exchange in (e). Now, star II explodes, and the entire system becomes a system consisting of two compact stars.

The evolution discussed above can explain why there are so few systems consisting of two compact stars in nature. During the

160

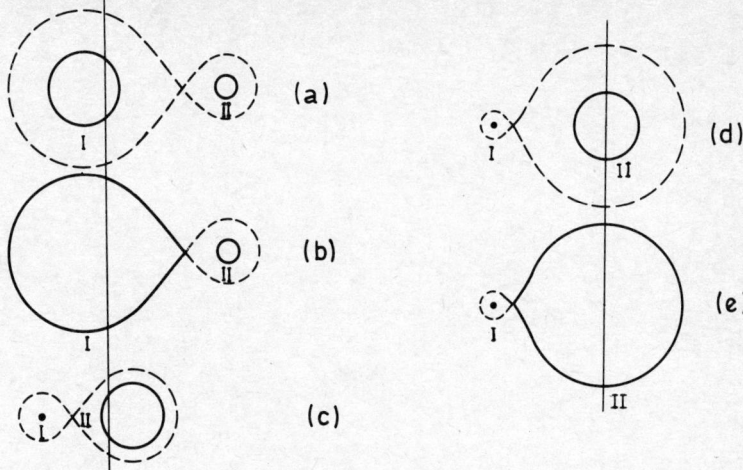

Fig. 5.4   Evolution stages of binary system

development from an ordinary star to a compact star, there is often a
supernova explosion which is likely to break up the binary system. The
evolution from an ordinary to a compact binary system has to undergo
two supernova explosions and it would be extremely rare to find a binary
system that has survived two such explosions.

It can be found, by comparing the evolution stages of various
kinds of binary stars that PSR1913+16 should be in the stage after
the second period of mass exchange, and consists of two compact stars.

In addition, the radiation of PSR1913+16 causes the rate of
change in rotation period to be

$$\frac{1}{T}\frac{dT}{dt} \sim \frac{16\pi R^2}{5c^2 p}\left(\frac{dp}{dt}\right) \sim 10^{-6} s/year$$

which is very small. Besides, the stellar wind, as mentioned previously, is not strong. All these properties indicate that PSR1913+16 is
a suitable system for testing the gravitational theory.

Starting in 1974, monitoring observations on PSR1913+16 were

161

maintained for 4 years. The following quantities were evaluated by analysing the time variation of pulse arrival:

$$\Omega = 4^\circ.226 + 0.002/\text{year},$$
$$\gamma = 0.0047 + 0.0007 \text{ s} ,$$
$$\sin i = 0.81 + 0.16 ,$$
$$dT/dt = (-3.2 + 0.6) \times 10^{-12} ,$$

where Ω is the angular velocity of the precession of the periastron, γ the second order Doppler factor and i is the inclination of the orbit plane. According to Newtonian mechanics and general relativity, there are relations between the above-mentioned quantities and masses m_1 and m_2 as follows:

$$\gamma = 0.002951 \left[\frac{m_1}{M_\odot} \right] \left[\frac{m_2 - 2m_1}{M_\odot} \right] \left[\frac{m_1 + m_2}{M_\odot} \right]^{-4/3} \text{ s} ,$$

$$\sin i = f^{1/3} m_1^{-1} (m_1 + m_2)^{2/3} = 0.5083 \left[\frac{m_1}{M_\odot} \right]^{-1} \left[\frac{m_1 + m_2}{M_\odot} \right]^{2/3} ,$$

where $f = 0.13 M_\odot$ called the mass function. Besides,

$$\frac{dT}{dt} = -\frac{96}{5} \frac{G^{5/3}}{c^5} T^{-5/3} \frac{M_\odot^{5/3}}{(2\pi)^{8/3}} f \left(\frac{m_1}{M_\odot} \right) \left(\frac{m_2}{M_\odot} \right) \left(\frac{m_1 + m_2}{M_\odot} \right)^{-1/3}$$

$$= -1.70 \times 10^{-12} \frac{m_1 m_2}{M_\odot^2} \left[\frac{m_1 + m_2}{M_\odot} \right]^{-1/3} ;$$

$$\Omega = \frac{3G^{2/3} M_\odot^{2/3}}{c^2 (1 - e^2)} \left(\frac{2\pi}{T} \right)^{5/3} \left[\frac{m_1 + m_2}{M_\odot} \right]^{2/3} = 2.11 \left[\frac{m_1 + m_2}{M_\odot} \right]^{2/3} \text{ deg s}^{-1}.$$

Thus, we have four observational quantities which contain two parameters m_1 and m_2 for evaluation. If the group of equations has self-consistent solutions, the theory will be proved to be reasonable. Since dT/dt is given by the theory of gravitational radiation damping, the observational results can provide a quantitative test for the theory.

In Fig. 5.5, the abcissa is m_1 and the ordinate is m_2. This includes the four above-mentioned relation curves of m_1 and m_2.

162

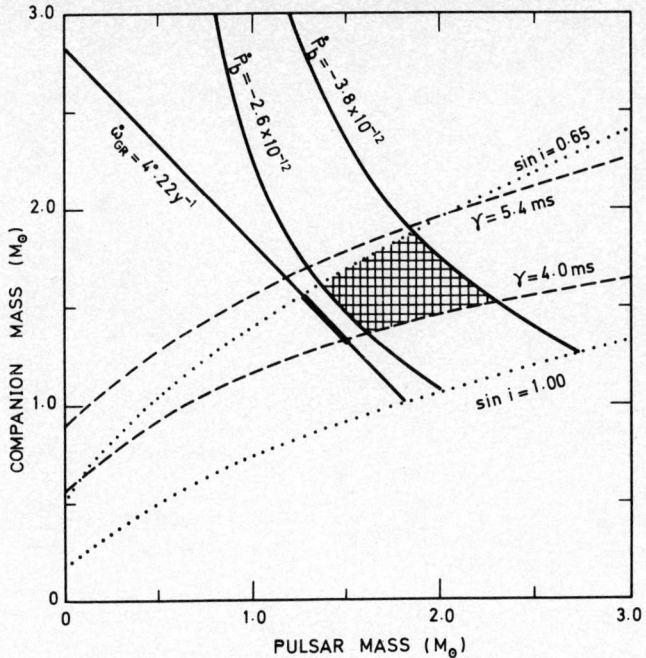

Fig. 5.5 γ, sin i and P_b obtained from observation.
The shadow expresses the most probable values
of the masses of the pulsar and its companion.

It is clear from Fig. 5.5 that the four curves have roughly the same
point of intersection. This quantitative evidence showing the
applicability of the theory of gravitational radiation damping. The
masses of the two stars can be determined by this figure to be
$m_1 \simeq m_2 \simeq 1.4\ M_\odot$.

This success can also be expressed in other ways. In Fig. 5.6
for example, the ordinate represents the orbit phase, and the abscissa
the observational time. If no gravitational radiation damping exists,
the phase line would be a straight line. Calculating the radiation
damping by general relativity and taking $m_1 \approx m_2 \approx 1.4\ M_\odot$, we can
get a curved phase line going downwards. The curved line from present
observational data coincides with theoretical predictions very well.

163

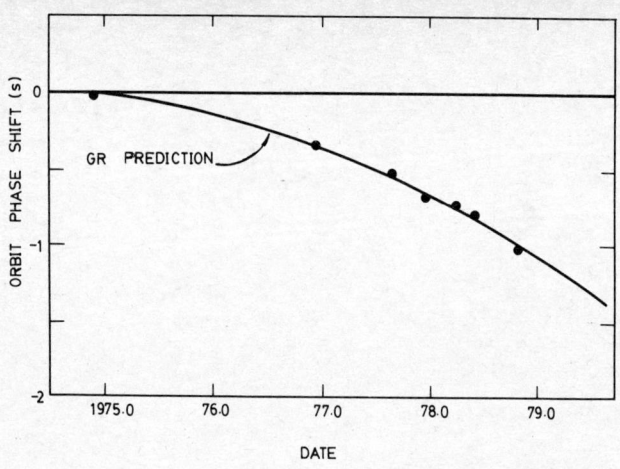

Fig. 5.6   The change of orbit phase of PSR1913+16
with respect to time

REFERENCES

Monographs on gravitational waves

I. Weber, General Relativity and Gravitational Waves (1961), (Inter-science publishers).

Concerning gravitational wave astronomy

W.H. Press, K.S. Thorne, Annual Review of Astronomy and Astrophysics, 10, 338 (1972).

Concerning the test of gravitational radiation damping

J.H. Taylor, L.A. Fowler, P.M. McCulloch, Nature, 277, 437 (1979).

Chapter 6

RELATIVISTIC COSMOLOGY

6-1 Difficulties between the Infinite Universe and Newtonian Theory

After establishing the general theory of relativity in 1916, Einstein promptly turned to the examination of the problems in cosmology. In 1917, he published his first paper on cosmology, "Kosmologishe Betrachtungen zur allgemeinen Relativitätstheorie", which was also to become the first paper on relativistic cosmology.

The reason why Einstein applied general relativity to cosmology so soon after the theory's introduction was because, at that time, cosmology represented the only field in which the significance of general relativity could be fully manifested. Among the various physical systems known then, only the universe had GM/c^2R near or equal to 1, which also meant that Newtonian theory was not applicable to the universe as a whole.

By the beginning of the 20th century, the correctness of Newtonian theory and the concept of an infinite universe seemed to have been confirmed. Indeed, the infinite universe as a concept of natural philosophy played an outstanding role in the struggle that broke the bonds of religiously sanctioned world views of the Middle Ages. The innovators, with Copernicus, Kepler and Galileo as representatives, applied ideas in the scientific spirit to overthrow the religiously approved doctrine of a geocentric universe. By the end of this scientific renaissance, Newton's theory had laid a firm dynamical foundation for the new concepts. The series of successes which followed facilitated the emergence and acceptance of a picture of the universe where space is infinite and unbounded whatever the direction taken. In other words, a structural view of the universe was accepted where universal space was considered as an infinite three-dimensional Euclidean space in which celestial bodies were distributed; but did this view rest on a true scientific basis? On the contrary, not only was there no real scientific foundation, there also existed certain intrinsic contradictions.

After every major event in science, which one would like to think of as progressive, the participants involved and others who arrive later have the common tendency not to examine certain key ideas more conscientiously, or else, perhaps owing to the tremendous prestige of such success, neglect to demand greater distinction between the areas covered by the newly advocated ideas which are based on established facts and those which still remain conjectural or inconclusive. This may result in various viewpoints being accepted without more stringent analysis. The infinite extent of Euclidean space and the validity of classical Newtonian theory in cosmology belong to the latter group of ideas to which many people have become accustomed to accepting when their basis is not clearly evident.

The initial step in Einstein's work was the analysis of the correctness of cosmological problems cast in these concepts. First of all, if the entire infinite universe is filled with matter, the gravitational field at infinity would not be zero. On the other hand, when we discuss local celestial motion, it is always assumed that we can choose a reference frame in which the gravitational field reduces to zero at infinity. This condition at infinite distance is necessary for resolving many problems in local celestial motion. We thus find ourselves in a plight in that we have either to abandon Newtonian mechanics as the dynamical basis of celestial motion, or accept the view that celestial

166

bodies are not uniformly distributed in the whole of infinite space, but instead occupy it on a finite scale.

Even if it is assumed that celestial bodies are found in a finite space, according to Newtonian theory, this system would not be stable; it would either collapse into itself or disintegrate when parts of it fly off.

When Newtonian theory is applied in the study of stars in an infinite space, the fundamental contradictions are unavoidable, and on such a basis the dynamics of an unbounded universe cannot, in principle, be established. Either Newtonian theory or the concept of an infinite universe or more drastically, both of them, need to be revised. This was one of the problems which Einstein had to face when he developed relativistic cosmology.

6-2   Spacetime of a Homogeneous and Isotropic Universe

According to general relativity, spacetime is curved from the very small atomic scale up to at least that of clusters of galaxies. In cosmology, one is concerned with structure on a large scale, ignoring it on small scales, that is, one imagines the inhomogeneities of spacetime as being averaged out over distances of the order of between 10^7 and 10^8 pc. This idea is equivalent to distributing matter in galaxies and clusters of galaxies uniformly throughout space. Such a procedure is often used elsewhere in physics. For example, in the study of non-turbulent flow in fluids, it is not, in general, necessary to consider the detailed motion of the atoms but rather their mean density and the total velocity of the fluid elements.

However, there are situations where this overall averaging process is not permitted in principle. The so-called hierarchical model of the universe insists on the impossibility of the above procedure. Indeed, the structure of the galaxies, clusters of galaxies and super-clusters indicate that the universe has hierarchical structure on some scale, but there is still insufficient observational evidence to compel us to accept such a model.

Therefore, we consider the universe as a cosmic fluid which is seen to be composed not of the ordinary atoms as the basic entities, but of galaxies and other celestial systems of smaller dimensions. We now set a clock for every fluid element where the clocks, called co-moving clocks, move with the fluid. The time indicated by a co-moving clock is the co-moving time, which is also the proper time of the fluid element. We wish to find out how the coordinate times given by different co-moving clocks on different elements are related.

To solve this problem, we propose a hypothetical cosmological principle: that every co-moving observer in the cosmic fluid will find the same history. By history, we mean the totality of physical phenomena in different epochs. It should be emphasised that the 'same history' is meant to be taken on the average, namely, it is not that the histories of the different fluid elements will be the same in detail, but that of averaged quantities on a large time scale. Under this assumption, we can synchronise all the clocks on different fluid elements. For example, all co-moving observers could set their clocks at the same time t when they see that the mean density ρ in their element has reached a certain value. Thus, we can use the time coordinate to describe the change of the universe with time.

The cosmological principle, which demands that the different parts of the cosmic fluid have the same histories, essentially states that the large scale universe is homogeneous. In the following discussion, we will also demand that the universe be isotropic, that is, each co-moving observer sees the same view in all directions in the cosmic fluid. This requirement is very strongly suggested by observations. For example, when the angular resolution is $1'$, the temperature of the microwave background radiation is independent of direction up to 10^{-3}. In addition, the distribution of radio sources seems to be isotropic as well.

Under the assumptions of homogeneity and isotropy, the three-dimensional space determined by $t = $ constant must be specified by a single curvature that must be the same at all positions, but may depend on time. We denote the constant curvature as $K(t)$, discussed in Chapter 1 and given by the formula (1.23).

We will now express (1.23) in terms of more convenient coordinates. We define

$$K(t) = \frac{k}{R^2(t)} \qquad \begin{cases} k = +1 & \text{positive curvature} \\ k = 0 & \text{flat space} \\ k = -1 & \text{negative curvature,} \end{cases}$$

where $R(t)$ is called the cosmic scale factor. We employ a new dimensionless coordinate χ defined by

$$\sqrt{K}\, r = \sin \chi \qquad 0 \leqslant \chi < \pi \qquad k = +1$$

$$\sqrt{K}\, r = \sinh \chi \qquad 0 \leqslant \chi \qquad k = -1 \quad .$$

Then, the spatial metric in (1.23) becomes

$$ds^2 = R^2(t)[d\chi^2 + \sin^2\chi(d\theta^2 + \sin^2\theta\, d\phi^2)]k = +1 \quad ,$$

$$ds^2 = R^2(t)[d\chi^2 + \sinh^2\chi(d\theta^2 + \sin^2\theta\, d\phi^2)]k = -1 \quad .$$

In these coordinates, the proper distance between any two points defined by fixed values of (χ,θ,ϕ) changes with time in the same way. Along the world line of a co-moving observer, where χ, θ and ϕ are constant, coordinate time t must be the same as proper time, and expressions for total spacetime separation would be

$$d\tau^2 = dt^2 - \frac{1}{c^2}R^2(t)\,[d\chi^2 + \sin^2\chi(d\theta^2 + \sin^2\theta\, d\phi^2)]k = +1 \quad ,$$

$$d\tau^2 = dt^2 - \frac{1}{c^2}R^2(t)\,[d\chi^2 + (d\chi^2 + \sin^2\theta\, d\phi^2)]k = 0 \quad ,$$

$$d\tau^2 = dt^2 - \frac{1}{c^2}R^2(t)\,[d\chi^2 + \sinh^2\chi(d\theta^2 + \sin^2\theta\, d\phi^2)]k = -1 \quad .$$

$$(6.1)$$

This is the Robertson-Walker metric first derived by Robertson and Walker in 1934, and it is only this metric which can satisfy the requirements of homogeneity and isotropy. It can be verified that χ, θ and ϕ as constants constitutes one of the sets of geodesics for which co-moving observers are in free-fall. It has to be emphasised that no specific gravitational arguments were employed in the derivation of the above metric, which instead is a consequence of the symmetry, homogeneity and isotropy of three-dimensional space.

At any time t, the spatial volume of the k = 1 universe is given by

$$V = R^3 \int_0^\pi \int_0^\pi \int_0^{2\pi} \sin^2\chi \, \sin\theta \, d\chi \, d\theta \, d\phi = 2\pi^2 R^3$$

which is finite, and therefore to be known as a closed universe. In the cases for which k = 0 and -1, the corresponding volumes are always infinite so that such universes are said to be open.

6-3   Relations between Apparent Magnitudes and Red-Shifts

Let us now discuss the observational evidence that would exist in a universe described by the Robertson-Walker metric, in order to be able to test the validity of this description by comparing predictions from theory with observations.

If a galaxy lies at the spatial origin $\chi = 0$ and another at χ, the proper distance between them at a given cosmic time t is given by (6.1) as

$$D_p = R(\tau)\chi \quad .$$

This distance is proportional to the cosmic scale factor R(t) and changes with time. If we define the proper velocity

$$V_p \equiv \mathring{D}_p \quad ,$$

we then have

$$V_p = \frac{\mathring{R}}{R} D_p \quad . \tag{6.2}$$

This tells us that at any given time t the relative speed of galaxies is proportional to the distance between them. If the proportional coefficient $\mathring{R}(t)/R(t) > 0$, one has a picture of expansive motion in the expansion of the universe.

It can be seen that although (6.2) may be used to indicate an

170

expanding universe, V_p and D_p cannot be measured directly. In order to verify the existence of an expansion, we have therefore to resort to other relations.

To prepare for this, we first analyse a simple problem. Consider light emitted at time t_e from a galaxy at $\chi = \chi_e$, reaching us at t_o. What change in wavelength would this light experience? Without losing generality, we can choose our spatial coordinate as $\chi = 0$. In Fig. 6.1, one wavelength of light arriving at t_o and $t_o + \Delta t_o$ was emitted at t_e and $t_e + \Delta t_e$. Since the light has travelled along a null geodesic, from the Robertson-Walker metric, the following relation would be satisfied:

$$dt^2 - R^2(t)\, d\chi^2/c^2 = 0 \quad , \tag{6.3}$$

or

$$c \int_{t_e}^{t_o} \frac{dt}{R(t)} = \chi_e$$

and

$$c \int_{t_e + \Delta t_e}^{t_o + \Delta t_o} \frac{dt}{R(t)} = \chi_e \quad .$$

Since Δt_e and Δt_o are much smaller than the period during which $R(t)$ changes, therefore, from the above formula we have

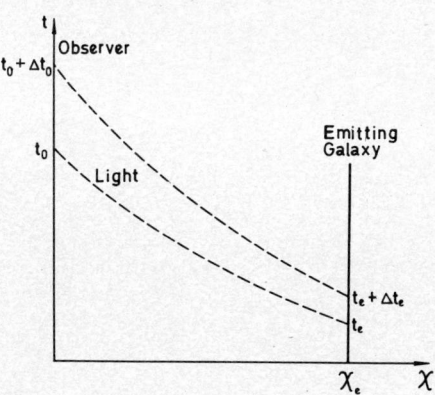

Fig. 6.1 Redshift in an expansive universe

171

$$\frac{\Delta t_0}{R(t_0)} - \frac{\Delta t_e}{R(t_e)} = 0 \quad ,$$

or

$$\frac{\Delta t_0}{\Delta t_e} = \frac{R(t_0)}{R(t_e)} \quad .$$

The observed and emitted wavelengths λ_0 and λ_e are related to the periods Δt_0 and Δt_e by

$$\lambda_0 = c \, \Delta t_0 \quad ; \qquad \lambda_e = c \, \Delta t_e \quad ,$$

so that

$$\frac{\lambda_0}{\lambda_e} = \frac{R(t_0)}{R(t_e)}$$

or, defining the redshift $z = \dfrac{\lambda_0 - \lambda_e}{\lambda_e}$ and using the above formula, we have

$$Z = \frac{R(t_0)}{R(t_e)} - 1 \quad . \tag{6.4}$$

In an expanding universe, $\dot{R}(t) > 0$, $R(t_0) > R(t_e)$, so that Z is positive, that is, all distant galaxies have redshifts. It is clear that (6.4) still cannot be compared directly with observations since t_e cannot be measured directly.

Most observed cosmological redshifts are rather small, so that t_e is not much earlier than t_0. Therefore, we can expand $R(t_e)$ as a Taylor series about $t_e = t_0$:

$$R(t_e) = R(t_0) + (t_e - t_0) \, \dot{R}(t_0) + \frac{1}{2} (t_e - t_0)^2 \, \ddot{R}(t_0) + \dots$$

$$= R(t_0) \left[1 + H_0(t_e - t_0) - \frac{1}{2} q_0 H_0^2 (t_e - t_0)^2 + \dots \right] \quad ,$$

where H_0 is the present value of Hubble's constant:

$$H_0 \equiv \dot{R}(t_0)/R(t_0) \quad , \tag{6.5}$$

and q_0 is the deceleration parameter, i.e.

$$q_0 \equiv - \ddot{R}(t_0)R(t_0)/\dot{R}^2(t_0) = - \ddot{R}(t_0)/R(t_0)H_0^2 \quad . \tag{6.6}$$

172

From these results, the redshift (6.4) may be expanded as

$$Z = H_0(t_0 - t_e) + (1 + \frac{1}{2}q_0) H_0^2(t_0 - t_e)^2 + \dots \quad .$$

The time interval $(t_0 - t_e)$ can be obtained from above power series inversion as

$$t_0 - t_e = \frac{1}{H_0}Z - (1 + \frac{1}{2}q_0)Z^2 + \dots \quad . \tag{6.7}$$

With this preparatory knowledge, we now analyse the problem on luminosity. Consider a celestial body whose absolute luminosity is L and whose distance from an observer is D_L, then, its apparent luminosity is

$$\ell = \frac{L}{4\pi D_L^2} \quad . \tag{6.8}$$

However, this formula takes no account of the curvature of space and the expansion of the universe. We need to know the form taken by this formula in the universe described by the Robertson-Walker metric.

Let the galaxy be at χ_e, and light emitted at t_e just reaches us at t_0. This light, at t_0, crosses the surface of a sphere whose area, from (6.1), is

$$A(\tau_0) = \begin{cases} 4\pi R^2(t_0) \sin^2\chi & k = +1 \\ 4\pi R^2(t_0) \chi_e^2 & k = 0 \\ 4\pi R^2(t_0) \sinh^2\chi_e & k = -1 \quad . \end{cases} \tag{6.9}$$

The power ℓ crossing unit area of this sphere is reduced by two effects in addition to the inverse-square law. First, because of the redshift, each photon arrives with its energy reduced by a factor $1 + Z$. Secondly, the rate of reception of photons is less than their rate of emission by the same factor $1 + Z$ for the same reason. The rate of emission of photons is related to the light frequency (the rate of emission of the number of waves), so that this 'frequency' would also have the same 'redshift'. Hence, the apparent luminosity in the Robertson-Walker metric, instead of (6.8), would be

$$\ell = \frac{L}{A(1 + Z)^2} \quad .$$

Using (6.4) for redshift, the above formula can be rewritten as

$$\ell = \frac{LR^2(t_e)}{AR^2(t_o)} \quad . \tag{6.10}$$

In order to obtain details for (6.10), we use the previous series expansion of (6.3) to give:

$$\frac{c}{R(t_o)} \int_{t_o}^{t_e} \left[1 + H_o(t_o - t) + (1 + \frac{q_o}{2}) H_o^2(t_o - t_e)^2 + .. \right] dt = \chi_e \quad ,$$

i.e.,

$$\chi_e \approx \frac{c}{R(t_o)} \left[(t_o - t_e) + \frac{1}{2} H_o(t_o - t_e)^2 + ... \right] .$$

Substituting (6.7) into this formula, we get

$$\chi_e \approx \frac{c}{R(t_o)} \left[(t_o - t_e) + \frac{1}{2} H_o(t_o - t_e)^2 + ... \right] ,$$

which when used in (6.10) with A from (6.9) gives

$$\ell = \frac{LH_o^2}{4\pi Z^2 c^2} \left[1 + (q_o - 1)Z + ... \right] . \tag{6.11}$$

Note that the relation between the bolometric apparent luminosity and the bolometric apparent magnitude m is

$$\ell = 10^{-2/5 \, m} \times 2.52 \times 10^{-5} \text{erg cm}^{-2} \text{s}^{-1} \quad .$$

Similarly, the relation between the bolometric absolute luminosity L and the bolometric absolute magnitude M is

$$L = 10^{-2/5 \, M} \times 3.02 \times 10^{35} \text{erg s}^{-1} \quad .$$

By substituting these relations into (6.11), we have

$$m - M = 25 - 5 \lg H_o + 5 \lg cZ + 1.086(1 - q_o)Z \quad , \tag{6.12}$$

which is the relation between apparent magnitude and redshift.

This significant result relates m and Z, which are directly measurable. However, M in the relation is not a measurable quantity which means that the determination of M is the greatest difficulty in verifying (6.12). To do this, the ordinary method would be to choose a group of celestial bodies of the same type whose absolute luminosity can be considered to be close. This group of celestial bodies would have M constant, and from (6.12), their m and Z would satisfy the following relation:

$$m = 5 \lg cZ + 1.086(1 - q_0)Z + \text{const.} \tag{6.13}$$

A similar relation between apparent magnitude and redshift for the colour magnitudes m_U, m_B, m_V can be also be derived. However, due to the redshift, m_U, m_B, m_V do not relate directly with M_U, M_B and M_V respectively, so that $m_U - K_U(Z)$, $m_B - K_B(Z)$ and $m_V - K_V(Z)$ should be used instead of m in (6.12) and (6.13) where $K_U(Z)$, $K_B(Z)$ and $K_V(Z)$ express the corrections, the so-called K corrections caused by redshift. The definite form of the K(Z) function depends on the radiation spectrum of the source.

For galaxies, Z is very small, so that the term containing q_0 in (6.12) and (6.13) can be generally neglected. In other words, a group of galaxies with same M would have

$$m = 5 \lg cZ + \text{const.}$$

or

$$m_V - K_V(Z) - A_V = 5 \lg cZ + \text{const.}$$

$$m_B - K_B(Z) - A_B = 5 \lg cZ + \text{const.}$$

$$m_U - K_U(Z) - A_U = 5 \lg cZ + \text{const.} \tag{6.14}$$

where A_U, A_B, A_V are the corrections caused by absorption within our galaxy and depend on galactic latitude.

After classifying various types of galaxies, it is indeed discovered that they all agree very well with (6.14). The relation between apparent magnitude and redshift for a group of typical galaxies is shown in the Fig. 6.2, generally known as the Hubble figure. This group of

175

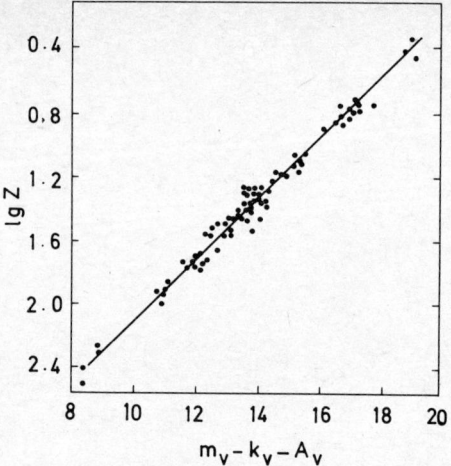

Fig. 6.2 Apparent magnitude redshift relation of
first-ranked cluster galaxies with
redshift of their parent clusters

galaxies consist of the brightest of 82 clusters of galaxies.

The constant in (6.14) can be determined by the Hubble figure obtained from observations. Comparing it with (6.12), we can obtain the Hubble constant. The values of the Hubble constant H_0 from different types of galaxy are quite close and are around

$$H_0 \approx 50 \sim 100 \text{ km s}^{-1}\text{Mpc} .$$

To determine the deceleration parameter q_0, we need to use celestial bodies with greater redshifts, because the term containing q_0 in (6.12) and (6.13) plays a role only when $Z > 0.2$. Obviously, it is very difficult to employ the apparent magnitude redshift relations of galaxies. The first problem is the lack of galaxies having large redshifts Z. Secondly, the method with which we select artificially the extraordinarily bright galaxies from different clusters of galaxies and consider that they have the same absolute magnitude (the so-called standard candle), is unreliable on the scale of large redshifts. The

176

larger the redshift, the more distant from us, so that light from a galaxy with a large redshift was emitted at a time when that galaxy was much younger than it is at the present cosmic time. Thus, the larger redshift indicates an earlier time. In this case, the question of galactic evolution must be considered, because the older galaxies might have had a very different absolute luminosity from present day galaxies. In particular, q_0 is highly sensitive to the uncertainty in the evolution correction, and this brings considerable uncertainty into the value of q_0.

These reasons compel us to turn to quasars for determining q_0 whose redshifts have quite a large spread. However, all quasars are distributed on the m_V - lg Z figure quite dispersively, which means that the crux of the issue is to choose the standard candle power. Obviously, it is not proper for quasars to take 'extraordinarily bright' as the standard candle without regard to their evolution. Since the life of a quasar, according to the general estimate, is shorter than that of a galaxy, the evolution effect has more notable influence. Therefore, the parameter for the standard candle for quasars should be an intrinsic physical quantity of a quasar, but a relative quantity among a group of quasars. Only in this way can the influence of evolution effects be evaded.

At present, several intrinsic physical quantities of quasars have been provided as the standard candle parameter, shown in Table 14.

177

Table 14.   Determinations of q_0 by quasars

Parameters of standard candle	q_0	Author
Steep radio spectrum	~ 1	Setti and Woltjer
Flat radio spectrum	~ 1	Stannard
L_α line width	~ 1	Davidsen *et al.*
Flat optical continuum	~ 1	Netzer *et al.*
Strong scintillation	~ 1.4	Qu *et al.*
Radio component separation	0.9 - 1.5	Fang *et al.*
CIV line	2.4 ± 1.4	Baldwin *et al.*
CIV, MgII, L_α line width	1.9	Kiang and Cheng

It is worth noticing that the values of q_0 between 1 ~ 2 are approximately the same for completely different standard candle objects.

6-4   The Darkness of the Night Sky

The darkeness of the night sky, i.e. Olbers' paradox, is well-known in classical cosmology. Now we apply the Robertson-Walker metric to discuss the problem of the intensity of the radiation background of the night sky.

According to the cosmological principle, the average luminosity $L(t)$ in space at any given cosmic time is the same everywhere. That $L(t)$ may change with time expresses the evolution of the galaxy. Let the number of these averaged galaxies in unit proper volume be $n(t)$, this being the proper number density. The proper volume between the hyperspheres χ, and $\chi + d\chi$ is, from the Robertson-Walker metric,

$$A(t)\ R(t)\ d\chi\ .$$

Thus, the number of galaxies enclosed in the volume is

$$n(t)\ A(t)\ R(t)\ d\chi\ .$$

If the galaxies are neither created nor destroyed, or if the created number equals the destroyed number, the above number remains constant. Using (6.9), this means that $n(t) R^3(t)$ is independent of time, because χ, and $\chi + d\chi$ are co-moving spheres, so that galaxies do not cross their surfaces on the average. Thus

$$n(t) R^3(t) = n(t_o) R^3(t_o) \quad ,$$

or

$$n(t) = \frac{R^3(t_o)}{R^3(t)} n(t_o) \quad . \tag{6.15}$$

The physical meaning of this formula is very clear: proper volumes are proportional to $R^3(t)$, so that densities decrease as $R^{-3}(t)$.

Using (6.10), we can calculate the present $(t = t_o)$ apparent luminosity of light emitted at t_e from all the galaxies between χ_e and $\chi_e + d\chi_e$ as follows:

$$d\ell = \frac{n(t_e) A(t_e) R(t_e) d\chi_e L R^2(t_e)}{A(t_o) R^2(t_o)} \quad .$$

Since light travels along null geodesics, we have

$$\frac{cdt_e}{R(t_e)} = d\chi_e \quad .$$

Thus, dt_e can be used instead of $d\chi_e$ in the expression for $d\ell$:

$$d\ell = \frac{cn(t_e) L(t_e) R^4(t_e) dt_e}{R^4(t_o)} \quad .$$

To find the total apparent luminosity, we integrate the above equation over the entire past history of the universe, from its origin t_{ori} up to the present time, giving

$$\ell = \frac{c}{R^4(t_o)} \int_{t_{ori}}^{t_o} dt_e \, n(t_e) L(t_o) R^4(t_e) \quad .$$

If the number of galaxies is conserved on the average, the above formula can be rewritten, by (6.15), as

179

$$\ell = \frac{n(t_0)}{R(t_0)} \int_{t_{ori}}^{t_0} dt_e \, L(t_e) \, R(t_e) \quad . \tag{6.16}$$

The night is dark, that is, the luminosity of the background is finite, so that ℓ certainly cannot be infinite. For 'big-bang' models with $t_{ori} = 0$, $R(t_{ori})$ is zero, and $L(t_e)$ is always finite. Thus the integral would converge, and ℓ is finite. This is, of course, a necessary but sufficient condition in order to avoid Olbers' problem, because the actual value of ℓ has not been given here. It must be less than the known intensity of the radiation background of every wavelength where such intensity in the X-ray and radio regions of the spectrum is very low.

For infinitely old universes, where $t_{ori} = -\infty$, we must have, for the integration in (6.16) to converge,

$$L(t) \, R(t) < \frac{constant}{|t|} \quad , \qquad \text{as} \quad t \to -\infty \quad .$$

In the classical model of the universe, where n, L and R are constants, we have

$$\ell = cnL \int_{-\infty}^{t} dt_e = \infty \quad .$$

This is the Olbers' paradox.

6-5 Number Counts

Using the method in section 6.4, we can calculate the number of galaxies in the entire celestial region which emitted light after t_e. This number, which we call $N(> t_e)$, is given by

$$N(> t_e) = \int_{t_e}^{t_0} n(t_e) \, A(t_e) \, R(t_e) \, \frac{d\chi_e}{dt_e} \, dt_e = c \int_{t_e}^{t_0} n(t_e) \, A(t_e) \, dt_e \quad .$$

If the number of galaxies is conserved, from (6.15), the above formula becomes

$$N(> t_e) = cn(t_0) \, R^3(t_0) \int_{t_e}^{t_0} \frac{A(t_e)}{R^3(t_e)} \, dt_e \quad . \tag{6.17}$$

180

By substituting the expansions for χ_e and $R(t_e)$, from section 6.3,

$$\chi_e \quad \frac{c}{R(t_0)} \left[(t_0 - t_e) + \frac{1}{2} H_0 (t_0 - t_e)^2 + \ldots \right] ,$$

$$R(t_e) = R(t_0) \left[1 + H_0 (t_e - t_0) + \ldots \right] ,$$

into (6.17) and integrating, we have

$$N(> t_e) = \frac{4\pi c^3}{3} n(t_0)(t_0 - t_e)^3 \left[1 + 3H_0 (t_0 - t_e) + \ldots \right] . \quad (6.18)$$

This result cannot be compared directly with observations yet, because t_e is not a directly measurable quantity. However, it can be found from (6.4) that galaxies emitting since time t_e have redshifts less than Z. Thus, in all galaxies in the entire celestial region, the number $N(< Z)$ of galaxies with redshifts less than Z is also the corresponding $N(> t_e)$. By applying (6.7), (6.8) becomes

$$N(< Z) = \frac{4\pi c^3 n(t_0) Z^3}{3 H_0^3} \left[1 + 3Z - 4(1 + \frac{1}{2} q_0) H_0 Z^2 \right] . \quad (6.19)$$

It is possible to compare directly this result on number counts of redshifts with observations. From all galaxies observed we can calculate $N(< Z) - Z$ which can be used, in principle, to determine the deceleration parameter q_0. However, there are difficulties obtaining reliable number counts in practice for the following reasons. The parameter $N(< Z)$ in (6.19) is the number of galaxies with redshifts less than Z, but every telescope has its limits and any celestial body with luminosity less than the minimum limit cannot be found, making it impossible to record all galaxies. This brings uncertainty in any comparison between theory and observation.

To overcome this difficulty in number counts, we can use the luminosity number counts $N(> \ell)$, which is to count the number of galaxies with apparent luminosity greater than ℓ. As long as ℓ exceeds the required observational minimum of the telescope, $N(> \ell)$ can be measured. Using (6.11), the expression for $N(> \ell)$ can be derived from (6.19) as

$$N(> \ell) = \frac{4\pi n(t_0)}{3} \left(\frac{L}{4\pi \ell}\right)^{3/2} \left[1 - \frac{3H_0}{c} \left(\frac{L}{4\pi \ell}\right)^{1/2} + \ldots \right] . \tag{6.20}$$

If the universe does not expand, $H_0 = 0$ and the above formula becomes

$$N(> \ell) = \frac{4\pi n(t_0)}{3} \left(\frac{L}{4\pi \ell}\right)^{3/2} .$$

This consequence is known as the $\ell^{-3/2}$ law in classical cosmology.
The influence of the expansion of the universe enables $N(> \ell)$ to be
smaller than the $\ell^{-3/2}$ law predicted. Equation (6.20) can be applied
to luminosity number counts not only for galaxies, but also for radio
sources. The result for radio sources is

$$N(> \ell) \simeq \ell^{-1.8} . \tag{6.21}$$

Obviously, this deviates from the $\ell^{-3/2}$ law, but the manner of departure is contrary to (6.20); there are more faint sources than what the
$\ell^{-3/2}$ law predicts.

What can we make of these results? It is obvious that one or
more of the approximations leading to (6.20) must be inapplicable.
Since observational results for the faint sources (small ℓ) is fainter
than theoretical predictions, there could be two reasons. The first is
that radio sources at an early time of the universe were, on the average, darker than that later. The second is that the number of radio
sources in the co-moving volume is not conserved, since in the past
they were more numerous than they are today. These two explanations
have the same conclusion: the universe evolves and is not in steady
state.

The very faint sources cannot be resolved by telescopes, and
their radio emissions form a continuous background. If (6.21) held for
these unresolved sources, this background would be much more intense
than it actually is. In fact the radio background is faint, so (6.21)
is inapplicable to fainter sources. This means that at an earlier
epoch of the universe the radio sources must have been, on the average,
fainter or less numerous than today. Observations on quasars support

182

this conclusion. As we know, quasars with redshifts $Z > 2$ are much rarer than those with $Z < 2$. This probably implies that at an earlier epoch, there were also fewer quasars.

To conclude from all this, we can speculate about the history of evolution of the universe. At an extremely early epoch there were no significant discrete sources of radiation until quasars, radio sources and galaxies began to form. In the beginning, these objects radiated abundantly but since that time, they have undergone gradual diminution.

6-6   Dynamical Equation for R(t)

The kinematics for a homogeneous and isotropic universe is completely described by the scale factor $R(t)$. $R(t)$ itself should be determined by the self-gravitation of all the matter in the universe, which means that the dynamical characteristics of $R(t)$ should be given by Einstein's field equations. Let us now find the dynamical equation for $R(t)$.

Firstly, all matter in the universe, with the fluid consisting of galaxies, is non-relativistic for the following reason: the ratio of pressure P to energy density ρc^2 is

$$\frac{P}{\rho c^2} \sim \frac{< v^2 >}{c^2} \, , \tag{6.22}$$

where $< v^2 >$ is the dispersive velocity of galaxies or clusters of galaxies. From the dynamics of galaxies or clusters of galaxies, the order of magnitude of $< v >$ is 10^2Nm s^{-1}, so that $P/\rho c^2 \sim 10^{-6}$, and pressure can hence be neglected. Matter in the universe is dust-like and is described by only one parameter, its mass density ρ.

The dynamical equation for $R(t)$ is just the relation between R and ρ. We shall derive this equation by using Newtonian mechanics rather than from Einstein's equation in detail.

Let us consider a sphere, marked by coordinate χ, with proper radius $\chi R(t)$. From homogeneity and isotropy, it follows that matter outside the sphere has a symmetric distribution. According to Newtonian gravitation, the outside matter does not act on the matter within the sphere. Therefore, the motion of inside matter is determined by their self-gravitational force. A galaxy on the surface of the sphere will be attracted by the mass within the sphere, where the total mass within the sphere is $4\pi/3 \ [\chi R(t)]^3 \rho(t)$. Thus, the gravitational force acting on the galaxy of mass m is

$$- Gm(\frac{4\pi}{3}) \ \rho(t) \ [\chi R(t)]^3 / [\chi R(t)]^2 \ . \tag{6.23}$$

The acceleration of the galaxy is $\chi \ddot{R}(t)$. Newton's second law gives the equation of motion for the galaxy as

$$m\chi\ddot{R} = - GM(\frac{4\pi}{3}) \ \rho(t) \ [\chi R(t)]^3 / [\chi R(t)]^2 \ ,$$

i.e.

$$\ddot{R}(t) = - (\frac{4\pi}{3}) \ G\rho(t) \ R(t) \ , \tag{6.24}$$

which is exactly the same as the equation from Einstein theory.

The reason why we can obtain the above result from Newtonian gravitation is the cosmological principle which tells us that the motions of the different parts of the universe are the same, so that we can study the dynamics of expansion of the universe in any very small local region. Relative speeds within such a small sphere are very much less than c, making Newtonian mechanics valid. Of course, if matter essentially involves large scale problems like redshifts, luminosities and non-Euclidean geometry, Newtonian mechanics is inapplicable.

In cosmology, we often put a new term in equation (6.24), which then becomes

$$\ddot{R}(t) = - 4\pi\rho(t) \ GR(t)/3 + \Lambda R(t)/3 \ . \tag{6.25}$$

Λ has dimension T^{-2} and is called the cosmological constant. From eq. (6.25), it can be found that Λ acts like a negative density ρ. Since the self-gravitation of positive ρ acts to slow down the expansion of the universe, a positive Λ must act to accelerate it and for this reason we call $1/3 \ \Lambda R(t)$ the cosmological repulsion term.

To solve (6.25), we first eliminate $\rho(t)$ using the law of conservation of matter. Since a co-moving volume is proportional to $R^3(t)$, the conservation law would be expressed as

$$\rho(t) \ R^3(t) = constant \ ,$$

or

$$\rho(t) = \frac{\rho(t_0) \ R^3(t_0)}{R^3(t)} \ .$$

Therefore, (6.25) can be rewritten as

$$\ddot{R}(t) = - \frac{4\pi\rho(t_0)R^3(t_0)G}{3R^2(t)} + \frac{\Lambda R(t)}{3} \ .$$

Multiplying by $2\dot{R}(t)$ and integrating, we obtain

$$\dot{R}^2(t) = \frac{8\pi\rho(t_0)R^3(t_0)G}{3R(t)} + \frac{\Lambda}{3} R^2(t) + const.$$

$$= \frac{8\pi}{3} G\rho(t)R^2(t) + \frac{\Lambda}{3} R^2(t) + const. \qquad .$$

The constant has dimensions L^2T^{-2}, and can be written as $-kc^2$. Thus, the above equation becomes

$$\dot{R}^2(t) = -kc^2 + \frac{1}{3} [8\pi G\rho(t) + \Lambda] \ R^2(t) \ . \qquad (6.26)$$

According to general relativity, k is not an arbitrary constant, but the constant in the Robertson-Walker metric has only one of three values: 1, 0 or -1. The value of k is a global property of the space-time of the universe, so that Newton's theory, which is valid only in the local region, is not sufficient to solve this problem.

Now, taking $t = t_0$ in the eqs. (6.25) and (6.26), and expressing $\dot{R}(t_0)$, $\ddot{R}(t_0)$ in terms of the Hubble constant H_0 and the deceleration parameter q_0, defined by (6.5), (6.6), we have

$$\Lambda = 4\pi\rho_0 G - 3q_0 \ H_0^2 \ , \qquad (6.27)$$

$$k = \frac{R^2(t_0)}{c^2} \ [4\pi G\rho_0 - H_0^2 \ (q_0 + 1)] \ , \qquad (6.28)$$

where $\rho_0 = \rho(t_0)$. The quantities ρ_0, q_0 and H_0 are measurable, so that Λ can be calculated from (6.27). In (6.28), it can be seen that the sign of k depends on the sign of the quantity in square brackets in (6.28). This means that the sign of $4\pi G\rho_0 - H_0^2(q_0 + 1)$ determines whether the universal space is infinite or not. In addition, knowning ρ_0, H_0 and q_0 from observations, (6.28) can be used to calculate $R(t_0)$.

If the cosmological constant $\Lambda = 0$, eq. (6.27) gives

$$q_0 = \frac{4\pi}{3} \frac{\rho_0 G}{H_0^2} = \frac{1}{2} \left(\frac{\rho_0}{\rho_c}\right) \quad , \tag{6.29}$$

where ρ_c is the critical density defined by

$$\rho_c \equiv \frac{3}{8\pi} \frac{H_0^2}{G} = 1.88 \times 10^{-29} h_0^2 \ g \ cm^{-3} \quad ,$$

and $h_0 = H_0/100 \ km \, s^{-1} Mpc$. Eliminating ρ_0 from (6.27) and (6.28), we obtain

$$k = 2 \left(\frac{R^2(t_0)}{c^2}\right) H_0^2 \ (q_0 - \frac{1}{2}) \quad . \tag{6.30}$$

It can be seen from this that whether the universe is closed or not depends on the deceleration parameter q_0 : $q_0 \geqslant 1/2$ indicates a closed universe, and $q_0 < 1/2$ indicates an open universe. Therefore $q_0 \sim 1-2$ given in section 6.3 suggests that the universe is probably closed.

On the other hand, when $q_0 > 1/2$, (6.29) demands

$$\rho_0 > \rho_c \quad .$$

However, the present value of ρ_0 obtained by observing galaxies is much smaller than ρ_c, and if this model is correct with a closed universe, there may exist much more invisible (undetected) matter, which would enable the actual ρ_0 to be much larger than the present known contribution of galaxies to ρ_0 in the universe. This can be considered as a 'missing mass' problem in modern cosmology.

186

6-7 Age of the Universe

Now let us solve the dynamical equation (6.26) of the scale factor $R(t)$. Using the mass conservation law $\rho(t)\,R^3(t) = $ constant, we
can rewrite (6.26) as

$$\dot{R}^2(t) = \frac{8\pi G\rho_0 R^3(t_0)}{3R(t)} - kc^2 + \frac{\Lambda}{3}R^3(t) \quad . \tag{6.31}$$

Now Λ, ρ_0, $R(t_0)$ and k can in principle be found from observations, as explained in the last section. Thus the parameters in (6.31)
are completely known, and we can obtain the solution for $R(t)$ and
hence the history of the universe in terms of $R(t)$.

Firstly, from the acceleration

$$\ddot{R}(t) = -\frac{4\pi\rho_0 R^3(t_0)G}{3R^2(t)} + \frac{\Lambda}{3}R(t) \quad ,$$

it follows that if $\Lambda = 0$ or $\Lambda < 0$, $\ddot{R}(t)$ is always negative, that
is, the expansion of the universe is always decelerated. Thus the
curve $R(t)$ is convex away from the t-axis (Fig. 6.3). At some time
in the past, $R(t)$ must therefore have been zero, and the density ρ
infinite; this is a singularity. We take $t = 0$ for the singularity
so that the present time t_0 is the age of the universe. If $\dot{R}$ is
constant, then

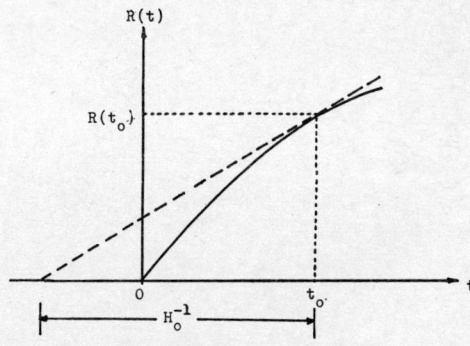

Fig. 6.3 Universe expanding from the 'big-bang'

187

$$\frac{dR}{dt} = \dot{R}(t) = R(t_o)H_o \quad .$$

Integrating the above formula and considering the condition $R = 0$ (when $t = 0$), we obtain

$$R(t) = R(t_o)H_o t \quad .$$

The age of the universe is $1/H_o$.

If the expansion is decelerated, $\dot{R}$ is not constant and is larger in the past than at the present time. The age of the universe in this model would be

$$t_o < 1/H_o \quad .$$

Now let us discuss the case of $\Lambda > 0$. From (6.26) it follows that when $k = 0$ or -1, $\dot{R}^2(t) > 0$, so that as long as $\dot{R}(t_o) > 0$, $\dot{R}$ is positive for all time. Therefore the universe must have begun from $R = 0$ and there would still be a singularity in this model. For $k = 1$ in (6.26), if

$$A(R) = \frac{8\pi G\rho_o R^3(t_o)}{3R(t)} - c^2 + \frac{\Lambda}{3} R^2(t)$$

is negative, then $\dot{R}$ can be zero and the singularity can be avoided. The minimum of $A(r)$ occurs at

$$R = \left(\frac{4\pi G\rho R^3(t_o)}{\Lambda} \right)^{1/3}$$

i.e.

$$[A(R)]_{min} \equiv (4\pi G\rho_o R^3(t_o))^{2/3} \Lambda^{1/3} - c^2$$

so that the condition for $A(R) < 0$ is

$$\Lambda < A_c \equiv \frac{c^6}{(4\pi G\rho_o R^3(t_o))^2} \quad .$$

Hence, based on (6.31), the only universes that need not begin from a singularity are those for which $k = +1$, $0 < \Lambda < A_c$. The qualitative form of $R(t)$ for this model is given in Fig. 6.4.

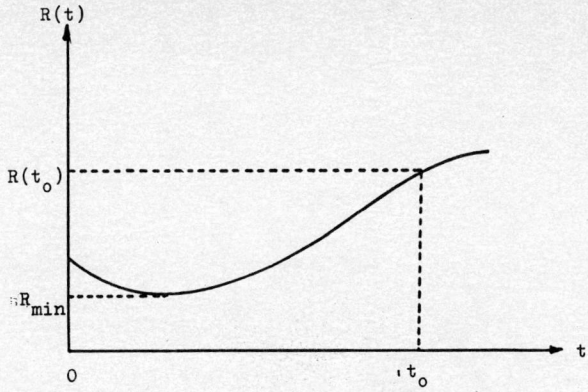

Fig. 6.4   Universe expanding after contraction to
finite 'radius'

Similar methods may be applied to examine the future of the universe.   If $\Lambda < 0$, $\ddot{R}$ is always negative, and the convex curve of
$R(t)$ must intersect the t-axis at some time in the future as well as
in the past.   Therefore, the present expansion must stop in the future
and be followed by a contraction, so that there will still be a singularity (Fig. 6.5).

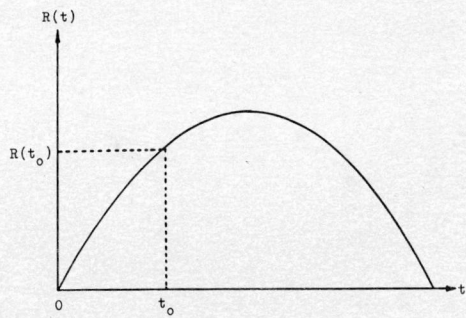

Fig. 6.5 $R(t)$ with singularities at both the
beginning and the future

189

If Λ is positive and $k = 0$ or -1, (6.31) tells us that $\dot{R}$ can never vanish, and the expansion continues forever. If $k = +1$, it is possible for $\dot{R}$ to vanish when $\Lambda < A_c$, for which the expansion may be followed by a contraction. The cases for $\Lambda = 0$ are that the ultimate collapse is avoided if $k = -1$ or 0, but not if $k = +1$.

These general properties of $R(t)$ can be summarised in Table 15.

Table 15. The properties of $R(t)$

		PAST	FUTURE
$\Lambda < 0$	$k = 0, \pm 1$	with singularity	with singularity
	$k = 0, -1$	with singularity	without singularity
$\Lambda > 0$	$k = 1$	with singularity without singularity $(\Lambda < A_c)$	without singularity $(\Lambda < A_c)$
$\Lambda = 0$	$k = 0, -1$ $k = 1$	with singularity with singularity	without singularity with singularity

Now we examine the rigorous solution for $R(t)$ with $\Lambda = 0$. In this case, all universes have a singularity in the past, that is, a 'big-bang' origin, no matter what k is. From eq. (6.31), we have:

$$k = 1, \quad R = \frac{R_m}{2}(1 - \cos \eta), \quad t = \frac{R_m}{2c}(\eta - \sin \eta) \quad ,$$

$$k = 0, \quad R = R_0 \left(\frac{t}{t_0}\right)^{2/3} \quad , \tag{6.32}$$

$$k = -1, \quad R = \frac{R_m}{2}(\cosh \eta - 1), \quad t = \frac{R_m}{2c}(\sinh \eta - \eta) \quad ,$$

where

190

$$R_m = \frac{8\pi G}{3c^2} \rho R^3 \, , \qquad\qquad (6.33)$$

(See Fig. 6.6). In the case of $k = 1$, the total mass of the universe is
$$M = 2\pi^2 \rho R^3 \, ,$$

and we thus have
$$R_m = \frac{4GM}{3\pi c^2} \, .$$

When $R = R_m$, $\dot{R} = 0$. Therefore, R_m is the maximum $R(t)$ of the closed universe. In other words, the closed universe expands to R_m, and then contracts, while the flat $(k = 0)$ and open $(k = -1)$ universes expand forever.

From eqs. (6.31) and (6.33), we have
$$R_m = \frac{2q_0 c}{|2q_0 - 1|^{3/2} H_0} \, .$$

Therefore, if $q_0 \sim 1$, as we saw in the last section, one finds
$$R_m \simeq \frac{2c}{H_0} = 0.6 \times 10^{10} \, h_0^{-1} pc \, .$$

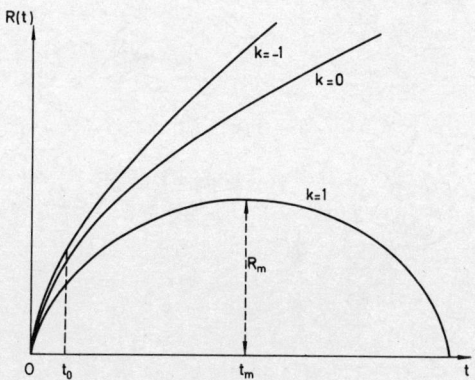

Fig. 6.6 Evolution history of different model universes with zero cosmological constant

191

The age t_m of the universe in its maximally expanded form is

$$t_m = \frac{\pi R m}{2c} = 3.07 \times 10^{10} h_o^{-1} \text{ years.}$$

An entire period between the two singularities in the past and in the future is

$$2t_m = 0.14 \times 10^{10} h_o^{-1} \text{ years.}$$

Setting $R = R(t_o)$ in (6.32), we can obtain the present age of these universes. For $k = +1$, we have

$$t_o = \frac{2q_o}{H_o |2q_o - 1|^{3/2}} \left[\sin^{-1} \sqrt{\frac{2q_o - 1}{2q_o}} - \frac{1}{2q_o} \sqrt{2q_o - 1} \right] \sim \frac{1}{6} t_m \quad .$$

If the space of the universe is flat, i.e. $k = 0$, this means that $q_o = 1/2$. Then, from (6.32) and (6.33), the present age would be

$$t_o = \frac{2R_m}{3} \left(\frac{R(t_o)}{R_m} \right)^{3/2} = \frac{2}{3H_o} = 0.6 \times 10^{10} h_o^{-1} \text{ years.}$$

Suppose that the universe is open, namely $q_o \sim 0$. Its age, from (6.32), would be

$$t_o = \frac{2q_o}{H_o |2q_o - 1|^{3/2}} \left[\frac{\sqrt{|2q_o - 1|}}{2q_o} - \sinh^{-1} \sqrt{\frac{|2q_o - 1|}{2q_o}} \right]$$

$$\sim \frac{1}{H_o} = 1.0 \times 10^{10} h_o^{-1} \text{ years.}$$

These values of age can be used to verify the corresponding models of the universe, because the various observed celestial bodies must form after the singularity, so that their age cannot be larger than t_o.

The first method for determining the age of celestial bodies is to use globular clusters, each of which is a group of stars formed at the same time. In their HR diagram, the turning point from the main sequence to red giants can be determined quite precisely. From the position of the turning point we can find the age of the globular

192

cluster. Using this method for many globular clusters, we have found that the age of the older globular clusters is about 9.0×10^9 — 1.0×10^{10} years. This result is comparable with the value of t_o given above.

The second method for determining age is to employ radioisotopes like ^{235}U, ^{238}U. From the study of the origin of elements, it is known that uranium originated from supernova explosions where the abundance ratio between ^{235}U and ^{238}U is 1.9. The abundance observed on the earth at the present time is 0.72% for ^{235}U and 99.27% for ^{238}U. From these two groups of data, it is deduced that the explosions which produced uranium should have taken place 7.0×10^9 years ago.

Similarly, the abundance ratios for ^{233}Th and ^{238}U, ^{187}Re and ^{187}Os, ^{129}I and ^{127}I can also be used for such analyses. Combining these results, we find that the solar system was born about 4.5×10^9 years ago, and the supernova explosions which could have produced the heavy elements in the solar system took place about 5.0×10^9 — 11.0×10^9 years ago. None of these values contradict calculations on the age of the universe.

6-8 Background Black-Body Radiation

In this section, we will discuss the properties of radiation in the expanding universe. Cosmic expansion causes the energy density $\varepsilon_r(t)$ of this radiation to decrease. On the one hand, the number density of photons decreases as $R^{-3}(t)$, while on the other hand, the energy of individual photons decreases as $R^{-1}(t)$ due to the redshift, with the net result that $\varepsilon_r(t)$ decreases as $R^{-4}(t)$, that is,

$$\varepsilon_r(t) = \frac{\varepsilon_r(t_o)R^4(t_o)}{R^4(t)} . \tag{6.34}$$

The equivalent mass density of the radiation is

$$\rho_r(t) = \frac{\varepsilon r(t)}{c^2} ,$$

which therefore also decreases as $R^{-4}(t)$.

According to (6.15), the mass density of matter in the universe decreases as $R^{-3}(t)$. As we go backwards in time, the radiation density increases faster than the matter density, so that although ρ_r is small now, there must have been a time when the densities of matter and radiation were equal. This means that when $t = t_E$, one had

$$\rho(t_E) = \rho_r(t_E) \ .$$

Since

$$\rho_0 = \frac{\rho(t_E)R^3(t_E)}{R^3(t_0)} \ ,$$

$$\rho_r(t_E) = \frac{\rho_r(t_E)R^4(t_E)}{R^4(t_0)} \ ,$$

we have,

$$\frac{\rho_0}{\rho_r(t_0)} = \frac{R(t_0)}{R(t_E)} \ . \tag{6.35}$$

At times before t_E, namely $0 < t < t_E$, ρ_r was greater than ρ. This period is called the radiation-dominated era of the universe.

As we shall see below, in the radiation-dominated era, temperature should have been higher and matter should have been in the plasma state, so that radiation and matter were in thermal equilibrium at the same temperature $T_r(t)$ because of the stronger coupling between them. Therefore, the radiation in equilibrium must have had a black-body spectrum. As the universe expanded, the mean photon energy decreased as $R^{-1}(t)$, so that the temperature should also have decreased as $R^{-1}(t)$. After t_E, i.e. $t > t_E$, the universe entered the matter-dominated era. As we shall discuss later, it so happens that at about $t = t_E$, radiation and matter ceased to be coupled in thermal equilibrium and the radiation temperature $T_r(t)$ thus needed no longer be equal to the matter temperature $T_m(t)$. No significant distortion of the black-body distribution occurred during the decoupling of matter and radiation. If the radiation preserves its black-body character after t_E, then $T_r(t)$ must continue to be proportional to $R^{-1}(t)$,

194

so that, if T_{ro} is the present radiation temperature, we have

$$T_r(t) = T_{ro} \frac{R(\overset{\bullet}{t}_0)}{R(t)} \quad . \tag{6.36}$$

Now, we shall prove that the radiation should indeed preserve its black-body character. The proof is based on the conservation of photon number. As we know, the photon number could have only been altered by interactions involving matter, but when $t > t_E$, the photons would have been decoupled from matter, and the number density n_r of photons would have been much larger than the number density n_m of elementary particles of matter. Such relatively few interactions could not have influenced the validity of the conservation of photon number.

According to Planck's black-body radiation law, the number $dN(t)$ of photons with frequencies between ν and $\nu + d\nu$ in a volume $V(t)$ is

$$dN(t) = \frac{8\pi\nu^2 V(t)d\nu}{c^3[\exp(h\nu/kT_r(t)) - 1]} \quad . \tag{6.37}$$

The number of photons in the co-moving volume $V(t)$ remains the same; in fact, there is a dynamic equilibrium which means that equal numbers of photons enter and leave the volume. At a new time t', the original group of photons would have been redshifted to a frequency

$$\nu' = \frac{\nu R(t)}{R(t')} \quad , \qquad \qquad d\nu' = d\nu \frac{R(t)}{R(t')} \quad ,$$

and the volume would have expanded to

$$V(t') = V(t) \frac{R^3(t')}{R^3(t)} \quad .$$

Therefore we have

195

$$dN(t') = dN(t)$$

$$= \frac{\frac{8\pi}{c^3}\left(\frac{\nu'R(t')}{R(t)}\right)^2 V(t')\, \frac{R^3(t)}{R^3(t')}\, d\nu'\, \frac{R(t')}{R(t)}}{\exp[h\nu'R(t')/R(t)kT_r(t)] - 1}$$

$$= \frac{8\pi V(t')\nu'^2 d\nu'}{c^3[\exp(h\nu'/kT_r(t')) - 1]} \quad ,$$

where

$$T_r(t') = T_r(t)R(t)/R(t') \quad .$$

Thus it has been proved that the radiation keeps its black-body distribution while the universe expands. The universe expands adiabatically and cools in accordance with (6.36). The energy density is obtained by setting $V(t) = 1$ in (6.37) and integrating:

$$\varepsilon_r(t) = \int dN\ h\nu = \frac{8\pi h}{c^3} \int_0^\infty d\nu\ \nu^3/[\exp(h\nu/kT_r(t)) - 1] \quad ,$$

i.e.

$$\varepsilon_r(t) = \frac{8\pi^5 k^4 T_r^4(t)}{15c^3 h^3} \equiv aT_r^4(t) \tag{6.38}$$

where $a = 7.5 \times 10^{-16}$ $Jm^{-3}K^{-4}$ is the Boltzmann-Stefan constant. Since T_r is an inverse measure of $R(t)$, so $\varepsilon_r(t)$ is an inverse measure of $R^4(t)$; this is perfectly consistent with (6.34).

In 1964, Penzias and Wilson first observed the microwave background radiation spread over all of cosmic space. At that time a horn-shaped antenna, which was originally designed to receive signals reflected by the Echo satellite, had been mounted in Bell Telephone Lab. They were prepared to measure radio emissions from high galactic latitudes. When they examined the installation on the waveband of 7.35 cm, they found unexpected noise, which could not be eliminated. The news was delivered to Princeton, and it was realized that the noise might come from the microwave background radiation in the cosmic space. This work published under the title "A measurement of excess antenna temperature at 4080 Mc/s" seemed to imply nothing of cosmology, but

196

it was to be the most important cosmological discovery since Hubble observed the redshift. A series of measurements were made on the waveband from 75 cm to 0.3 cm. Measurements on wavelengths larger than 100 cm failed, because the high-frequency radiation from our galaxy itself conceals the extragalactic radiation. The waveband of less than 3 mm lies outside the range for which the atmosphere is a transparent window and the corresponding measurements must rely on balloons or rockets. The summarized result shown in Fig. 6.7 indicates that this curve fits very closely to a black-body distribution with temperature $T_{ro} = 2.7$ K. This means that most of the radiation has wavelengths of the order of millimetres. From (6.38), we obtain the present value of the energy density of cosmic radiation:

$$\varepsilon_r(t_0) = aT_{ro}^4 = 4 \times 10^{-13} \text{erg cm}^{-3} \ , \tag{6.39}$$

that is, $\rho_r(t_0)$ is about a thousandth of the mass density $\rho_m(t_0)$ of matter. Thus the universe is indeed matter-dominated at the present time. Of course, the universe does not just contain microwaves but also all other kinds of radiation. However, the background radiation of

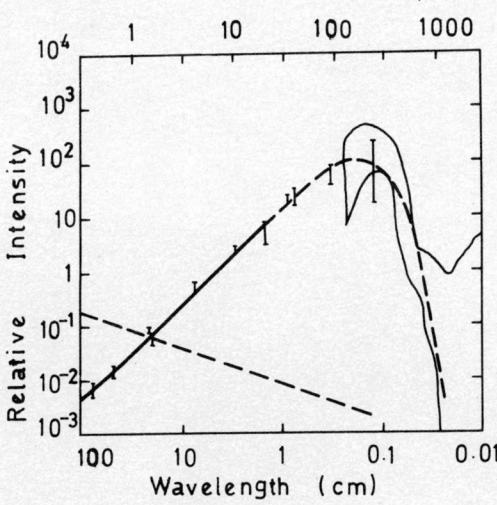

Fig. 6.7   Intensity of cosmic background radiation

197

all other wavelengths is very weak and their total energy density has been estimated to be less than $\varepsilon_r(t_o)/100$. Morever, these radiations are not distributed with a black-body spectrum. Microwave background radiation therefore has the most important cosmological significance. The microwave background radiation may be perfectly isotropic on both the small and the large-scale.

A large radio antenna can be used for observing the small-scale isotropy (the larger the antenna, the higher the resolution) with the antenna fixed at a certain angle relative to the earth. The direction of the antenna would thus scan different regions of the celestial sphere. We can determine the small-scale anisotropy due to small fluc- tuations of the temperature of the antenna. Several observational results are listed in Table 16 below.

Table 16. The small-scale anisotropy of the background radiation

Wavelength (cm)	Angular scales	Relative fluctuation
7.35	40'	0.2 %
3.95	1.4' X 20'	0.03 %
2.80	10'	0.2 %
0.35	80'	0.2 %

From the table, it can be seen that the fluctuation of radiation intensity, on scales less than tens of minutes, is still lower than two or three thousandth. It is therefore obvious that there is small- scale isotropy.

As for measuring the large-scale isotropy, we can use a horn- shaped antenna for a certain interval of time to measure alternately the radiation arriving from the different directions of the celestial equator and the north pole. According to the change in difference between two signals with time, we can obtain the intensity of back- ground radiation for different directions. The diurnal rotation of the Earth around its axis introduces a succession of different directions

198

every day and hence, a period of 24 hours for the observed radiation if one assumes the existence of an intrinsic anisotropy of the field. Analysing the component with a period of 24 hours from the differences between two kinds of signals, we can measure the degree of large-scale anisotropy. The results of some recent measurements are listed in Table 17.

Table 17. The large-scale anisotropy of the background radiation

Wavelength (cm)	Period (hour)	Relative change
3.75	24	0.06 %
3.2	12	0.06 %
3.2	24	0.04 %
0.8	12	0.2 %
0.8	24	0.28 %

It follows that quite high isotropy also exists on the large-scale.

These results imply that microwave background radiation may be the most concrete embodiment of co-moving coordinates in the homogeneous and isotropic model of the universe, which means that co-moving coordinates are set in a homogeneous and isotropic background radiation. In other words, on a local scale, the coordinate system taking the background radiation as reference is the best inertial system. According to this view of point, we can measure the motion of the Earth or the Sun relative to the co-moving coordinate system, i.e. to the background radiation. If this relative speed equals zero, the temperature T_o of the background radiation would be exactly isotropic. If the relative speed has a non-zero value v, by the Doppler effect, the temperature of radiation from the forward direction would be higher, and temperature in the backward direction would be lower. The temperature depends on direction by the following relation:

$$T(\theta) = T_o (1 - \frac{v^2}{c^2})^{1/2} (1 - \frac{v}{c} \cos \theta)^{-1} \quad ,$$

where θ is the polar angle with the direction of motion as axis. The most recent measurement gives

$$v_{\odot} = 390 \pm 60 \text{ km s}^{-1} \text{ ,}$$
$$v_{GL} = 600 \text{ km s}^{-1} \text{ .}$$

Now we study the transition between the radiation and matter-dominated eras. Substituting (6.39) and $\rho_o \sim \rho_c \simeq 1.9 \times 10^{-29} \text{g cm}^{-3}$ into (6.35), we have

$$\frac{R(t_o)}{R(t_E)} \approx 4.2 \times 10^4 \text{ .}$$

The transition temperature is about

$$T_r(t_E) \simeq 1.1 \times 10^5 \text{ K} \text{ .}$$

When the temperature is above 4000 K, it is known that hydrogen would be completely ionised. Thus, at the transition time $t = t_E$, free electrons and protons should recombine into neutral hydrogen, and since the plasma interacts more strongly with radiation, before the recombination, radiation and matter should be in thermal equilibrium. After the recombination, the matter should have become transparent to radiation and the coupling between radiation and matter would have disappeared.

In the radiation-dominated era, for small t, $R(t)$ is also small and $\rho(t)$ is large. Thus, in the dynamical equation (6.26), the terms involving the curvature index k and the cosmological constant are negligible. We thus have, for different expanding models, the same dynamical equation as follows:

$$\dot{R}^2(t) = \frac{8\pi}{3} G\rho(t)R^2(t) \text{ ,}$$

or

$$(\frac{dR^2}{dt})^2 = \frac{32\pi}{3} G\rho R^4 \text{ .}$$

In principle, ρ must be due to the total contribution of both radiation and matter. For $t < t_E$, ρ_r dominates, so that $\rho \sim \rho_r$. Since

200

$\rho_r R^4$ = constant, the solution of the above equation with initial condition $R(0) = 0$ is

$$R = (\frac{32\pi}{3}G\rho_r R^4)^{1/4} t^{1/2} \quad ,$$

i.e.

$$\rho_r = \frac{3}{32\pi G t^2} \quad . \tag{6.40}$$

ρ_r can be expressed as

$$\rho_r = g \frac{\pi^2}{3c} (\frac{T}{\hbar c})^3 T \tag{6.41}$$

where $g = g_b + \frac{7}{8} g_f$, and g_b and g_f are the number of particle spin states of bosons and fermions respectively. For photons, $g_b = 2$, $g_f = 0$. Using (6.40) and (6.41) one finds

$$T^4 = \frac{45}{16\pi^3 g} \frac{(\hbar c)^3}{G} \frac{1}{t^2} \quad . \tag{6.42}$$

It should be emphasised that these formulae apply irrespective of the values of Λ and k, and contain no adjustable constants. Of course, it is not strictly correct to apply these formulae to the transition era t_E, because the matter content is then no longer negligible, and in addition the terms involving Λ and k might have been significant. However, if we are interested only in rough estimates, we can still use (6.40) and (6.41), and we find that the value for recombination time is

$$t_R = \left(\frac{3c^2}{32\pi Ga} \right)^{1/2} \frac{1}{(4000 \text{ K})^2} \simeq 10^5 \text{ years} \quad .$$

Therefore, the radiation-dominated era is a very short period in the history of the universe when it was radiation-dominated.

The question now is how to verify these conjectures about the early universe. In the next section, we will discuss this problem.

6-9   Helium Abundance

If the early universe was very hot, and matter and radiation were in thermal equilibrium, only elementary particles were present at that time, such as protons, neutrons, electrons, photons, neutrinos and their anti-particles. Nuclei and atoms were gradually formed only after the cooling of the expanding universe.

During the period when the cosmic temperature was lower than 10^{12}K, the equilibrium between neutrons and protons in the universe was maintained by the following reactions:

$$\nu_e + n \leftrightarrow p + e^- \ ,$$

$$e^+ + n \leftrightarrow p + \bar{\nu}_e \ .$$

The rates of these reactions are approximately equal with weak processes as follows:

$$e^+ + e^- \leftrightarrow \nu_e + \bar{\nu}_e \ ,$$

$$e^\pm + \nu_e \leftrightarrow e^\pm + \nu_e \ ,$$

$$e^\pm + \bar{\nu}_e \leftrightarrow e^\pm + \bar{\nu}_e \ .$$

Since the expansion of the universe should have destroyed the thermal equilibrium, the equilibrium condition was that the rates of the reactions be larger than that of the expansion of the universe. For the above reactions, the equilibrium condition was no longer satisfied when the cosmic temperature was lower than $T_d = 10^{10}$K. Hence, when $T < T_d$, the neutron-proton ratio was frozen at the thermal equilibrium value at T_d.

$$n/p \sim \exp\left(-\frac{\Delta mc^2}{kT_d}\right)$$

where Δm is the mass difference between the neutron and the proton. The value of n/p is about $1/6$.

When cosmic temperature $T < T_d$, there were two possibilities for the evolution of neutrons: one was free decay and the other was to merge with protons to form nuclei.

The nucleosynthesis processes can begin only when the cosmic temperature drops to $T_s \sim 10^9$K, known as the nucleosynthesis temperature. This is because the first step in element formation is the

202

building of deuterium with its very low binding energy of 2.22 MeV.
When $T > 10^9$ K, deuterium will be broken down in the reaction
$\gamma + D \rightarrow p + n$. When $T < 10^9$ K, high energy photons were few and
deuterium formed rapidly. In the next step, deuterium is synthesised
rapidly into helium and other nuclei with larger mass by the following
reactions:

$$D + p \rightarrow {}^3He + \gamma \qquad ,$$

$${}^3He + {}^3He \rightarrow {}^4He + 2H + \gamma \quad ,$$

$${}^3He + {}^4He \rightarrow {}^7Be + \gamma \qquad ,$$

$${}^7Be + e^- \rightarrow {}^7Li + \nu \qquad .$$

It can be said that almost all neutrons at T_s would be bound
into ${}^4He$ with a very small percentage in D, ${}^3He$, ${}^7Li$ etc.
Between T_d and T_s, some neutrons disappear by free decay, so that
the ratio n/p drops from 1/6 at T_d to 1/7 at T_s. The mass
fraction of ${}^4He$ formed in the early universe is then

$$Y = \frac{2n}{p + n} \Big|_{T = T_s} \simeq 0.25 \quad .$$

The abundance of ${}^4He$ in the early universe cannot, in general,
be tested by present observations, because these abundances were modi-
fied by nuclear processes in stars after the early stage of the universe. However, the amount of ${}^4He$ produced in the stellar processes
is very small and the primordial value of ${}^4He$ abundance should
approximately be the present observed value. Table 18 shows the
observational results.

The measure of element content in celestial bodies is based
mainly on relative intensities of spectral lines. There are ionic
hydrogen regions with high temperatures in every galaxy, so that it is
usually possible to observe the emission lines of hydrogen and helium.

The temperature on the Sun's surface is low, only about 6000 K,
so that the spectral lines of He cannot be observed. However, when
solar prominences appear, the result $Y \simeq 0.38$ has been estimated from

203

Table 18. ^{4}He abundance Y

Observed objects	Result of Y
HII region	0.32 ± 0.01
Orion nebula	0.28 ± 0.01
LMC and SME	0.25 ± 0.01
young galaxies	0.23 ± 0.02
underluminous galaxies	0.24 ± 0.01
emission-line dwarf galaxies	0.25 ± $\begin{matrix} 0.01 \\ 0.02 \end{matrix}$

the spectrum of solar prominences. The observation of solar wind can also be used for estimating $\overline{Y}$, found to be $\overline{Y} \simeq 0.20$. It follows that all these direct measurements, which have given $\overline{Y}$ as about 0.3 or so, coincides quantitatively with theoretical values.

All of the measurements mentioned indicate directly, in fact, only the abundance on star surfaces. The abundance of stellar interiors cannot be measured directly, but only inferred indirectly by using the HR diagram for globular clusters. If the hydrogen abundance X, helium abundance Y and abundance Z of all other elements are given, then the curve on the HR diagram can be obtained from the theory on stellar structure. By comparing this curve with the observational curve, the value of Y, which is between 0.24-0.33, has been obtained; this coincides well with the universal value from spectral observations.

This method can also be used for single stars: if the mass, luminosity, age and heavy elements abundance Z of a star are known, the stellar structure theory could give Y. For instance, for the Sun, it is known that $M_{\odot} = 1.989 \times 10^{33}$ g, $L = 3.83 \times 10^{33}$ erg s^{-1}, and the relative abundance of heavy elements to hydrogen is $Z/X \simeq 0.019$, as is estimated from the absorption spectra of hydrogen and heavy elements, and an age which is estimated to be about 4.5×10^9 years as mentioned before. From all this, we find that the helium abundance of the Sun is also about 0.27-0.30. It seems that the theory agrees

204

very well with observations once again.

6-10　Formation of Galaxies

So far we have explored the uniform universe in which radiation and matter are homogeneously distributed. This assumption of uniformity might hold on the large scale but not on the small scale. On a scale less than about 100 Mpc, matter exists in the form of galaxies, clusters of galaxies and superclusters. In this section we will discuss the formation of such inhomogeneities.

In order to answer this question, it is obviously necessary to know what inhomogeneities could have existed in the primordial stage of the universe. One model holds that there were not only large scale homogeneities but also small scale ones in the early universe. However, the model has met difficulties, connected mainly with the so-called causality region of early universe.

Consider light emitted from $\chi = 0$ at $t = 0$ and reaching χ at t. According to the propagation equation of light, we have

$$\chi = c \int_0^t \frac{at}{R(t)} \quad .$$

Taking into account the early solution (6.40) of ρ_r and the relation $\rho_r R^4 = \text{constant}$, one finds

$$\chi \propto \sqrt{t} \quad ,$$

or

$$R\chi \sim ct \quad .$$

This is the size of regions in which causality holds. It will be small when t is small, so that there is obviously no physical basis for assuming that regions of the universe which could never have been causally connected should have the same density. For these reasons, a more reasonable model should accept the assumption that there were initial density inhomogeneities on all scales.

How do these inhomogeneities develop? Basically, three processes

205

govern their development: self-gravitation enhances inhomogeneities, the pressure from matter, and radiation as a source of oscillations which causes an inhomogeneity to radiate in the form of waves. Photon viscosity is a source of damping, causing an inhomogeneity to decay. It arises from collisions between photons and particles of matter.

Let us first analyse the mechanism of gravitational instability. If there is an initial inhomogeneity with radius r in a fluid of density ρ_T, the total mass in this region is about $M \sim \rho_T r^3$, the characteristic gravitational speed is $\sqrt{GM/r}$, and the characteristic time of self-gravitational contraction is then

$$\frac{r}{\sqrt{\frac{GM}{r}}} \sim (G\rho_T)^{-1/2} \ .$$

On the other hand, the pressure enables the inhomogeneity to produce sound waves. If the speed of the sound is v, the characteristic time for this pressure-related process is r/v. Thus, if

$$(G\rho_T)^{-1/2} > \frac{r}{v} \ ,$$

self-gravitation dominates and gravitational instability is present.

The above condition can also be written as

$$M_T = \frac{4\pi}{3} \rho_T r^3 > M_{TJ} \equiv \frac{4\pi}{3} v^3 G^{-3/2} \rho_T^{-1/2} \tag{6.43}$$

where M_T is the total mass, ρ_T the total density, M_{TJ} is called Jeans mass. This formula indicates that if M_T exceeds M_{TJ}, the inhomogeneity will develop into a condensed celestial body. If M_T is less than M_{TJ}, the density will oscillate like a sound wave.

In cosmological problems, we are more interested in the mass M of matter than the total mass M_T. Analogous to the preceding derivation, the condition (6.43) should be rewritten as follows:

$$M > M_J \equiv \frac{4}{3}\pi\rho v^3 (G\rho_T)^{-3/2} \ , \tag{6.44}$$

where ρ is the mass density of matter, $\rho_T = \rho + \rho_r$ is the total mass density, and ρ_r is the radiation density.

Now, we use these conditions to discuss the development of initial inhomogeneities in the expanding universe.

1. Before the recombination era $(t < t_R)$, the matter density is negligible, we have

$$\rho_T \simeq \rho_r = \frac{aT_r^4}{c^2} .$$

The equation of state is

$$p = \frac{1}{3} \rho_r c^2 .$$

According to thermodynamics, the speed of sound is

$$v = \sqrt{\frac{dp}{d\rho_T}}$$

so we have

$$v = \frac{c}{\sqrt{3}} ,$$

and the Jeans mass is

$$M_J = \frac{4\pi\rho c^3}{3(3GaT_r^4/c^2)^{3/2}} \approx \frac{\rho c^6}{T_r^6 (Ga)^{3/2}} .$$

To find the matter density ρ in this era, we use (6.35) and (6.36) to get

$$\rho(t) = \rho(t_E) \frac{R^3(t_E)}{R^3(t)} = \frac{\rho_r(t_E)T_r^3(t)}{T_r^3(t_E)} = \frac{aT_r(t_E)T_r^3(t)}{c^2}$$

that is,

$$M_J^{(t)} = \frac{c^4 T_r(t_E)}{G^{3/2} a^{1/2} T_r^3(t)} .$$

207

It follows that as the universe expands and the radiation cools, M_J increases by T_r^{-3}. If $t_E \sim t_R$, i.e. $T_r(t_E) \sim T_r(t_R)$, where t_R denotes the recombination time, we then have

$$M_J(t_R) \simeq 10^{18} M_\odot \quad ,$$

where we have taken $T_r(t_R) \sim 4000$ K. This value of M_J greatly exceeds typical galactic masses. Thus an initial inhomogeneity with galactic mass will contract only after the recombination. If M_J exceeds the typical galactic mass, the initial inhomogeneity with scale M_J should disintegrate and become a sound wave.

2. After the recombination era $(t > t_R)$, the radiation density is negligible, and $\rho_T \simeq \rho$. Since matter would then consist mainly of hydrogen and helium, we can treat it as an ideal gas, with specific heat ratio $\gamma = 5/3$ and the equation of state

$$p = \frac{\rho K T_m}{m} \quad ,$$

where m is the atomic mass of the gas, and T_m the matter temperature, which is not the same as T_r. Thus, the speed of sound is given by

$$v = \frac{5}{3} \frac{k T_m}{m} \quad .$$

From (6.44), the Jeans mass is

$$M_J = \frac{4\pi}{3} \left(\frac{5}{3} \frac{k T_m}{Gm} \right)^{3/2} \frac{1}{\rho^{1/2}} \quad . \tag{6.45}$$

At the recombination time t_R, $T_m = T_r$ and

$$\rho = \frac{\rho_0 R^3(t_0)}{R^3(t_R)} = \frac{\rho_0 T_r(t_0)}{T_r(t_R)} \quad .$$

From (6.45) we then have

208

$$M_J(t_R) \simeq 1 \times 10^5 \left(\frac{\rho_0}{\rho_c}\right)^{-1/2} M_\odot \quad .$$

After the recombination, the expansion of matter in the universe is adiabatic, namely

$$T_m \propto V^{-(\gamma-1)} = V^{-2/3}$$

or

$$T_m \propto R^{-2} \propto T_r^2 \quad .$$

In addition, $\rho \propto R^{-3} \propto T_r^3$, and (6.45) can hence be written as

$$M = M_J(t_R) \left(\frac{T_r(t)}{T_r(t_R)}\right)^{3/2} \quad .$$

This means that as the universe expands and cools, M_T decreases as $T_r^{3/2}$. After the recombination, M_T is always much less than a typical galactic mass, so that any inhomogeneity with galactic mass would develop.

The behaviour of the Jeans mass as a function of T_r is shown in Fig. 6.8.

We now consider the effect due to photon viscosity. Before t_E, photons collide frequently with electrons. In the presence of inhomogeneities, the diffuse photons cause the migration of matter and the smooth of matter inhomogeneities. If L denotes the photon mean free path, the total number of photon collisions at the universal age t would be $N = ct/L$. Because diffusion is a random walk process, the distance diffused by a photon in the time t would be $L\sqrt{N}$. Thus, all inhomogeneities smaller than the distance will decay and we assume that all inhomogeneities with size X smaller than $1/5\,L\sqrt{N}$ will have been damped out, that is,

$$X(t) \quad \frac{1}{5} L \sqrt{N} \quad = \frac{1}{5} L \sqrt{\frac{ct}{L}} = \frac{1}{5} \sqrt{Lct} \quad .$$

From kinetic theory,

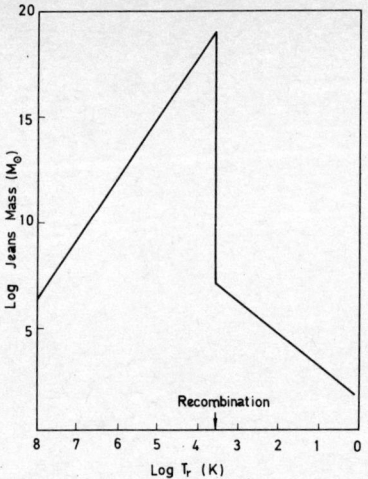

Fig. 6.8   Relation between the Jeans mass and the
radiation temperature T_r where the solid
line corresponds to T_{ro} = 2.7 K,
ρ_r = 3 X 10^{-29}g cm^{-3},   the dotted line
corresponds to T_{ro} = 2.7 K, ρ_0 = 10^{-30}g cm^{-3}.

$$ L = \frac{1}{n_e(t)\sigma} = \frac{m}{\rho\sigma} \ , $$

where n_e is the number density of scatterers such as electrons, and
σ is the scattering cross-section, which for photon-electron collisions has the Thomson cross-section

$$ \sigma = 6.65 \text{ X } 10^{-24} \text{ cm}^2 \ . $$

Thus the mass $M_D(t)$ of the smallest inhomogeneity that survives undamped is

$$ M_D(t) = \frac{4\pi\rho(t)}{3} \left(\frac{X(t)}{2}\right)^3 = \frac{4\pi}{3 \text{ X } 10^3 \rho^{1/2}(t)} \left(\frac{mct}{\sigma}\right)^{3/2} \ . $$

The minimum undamped mass before the recombination is

$$ M_D(t) = \frac{4\pi}{3 \text{ X } 10^3 \text{ X } \rho^{1/2}(t_R)} \left(\frac{mct_R}{\sigma}\right)^{3/2} \approx 10^{11} \ \frac{\rho_0}{\rho_c}^{-1/2} M_\odot \ . $$

210

This value is very close to the typical galactic mass.

So far, the several mechanisms only give us the characteristic masses, which are not in contradiction with the mass that is necessary for galaxies to form. Why are the galactic masses close to $10^{12}M_\odot$? How do galaxies contract to the size observed at the present time? These are still open questions.

Another open question concerns the origin of galactic angular momentum. Many galaxies are disc-shaped, which results from galactic angular momentum. As a galaxy contracts under its self-gravitation, its angular momentum is conserved, that is, there is a centrifugal potential resisting contraction. For this reason, contracting matter only moves in a circular orbit with an axis along the direction of the angular momentum. However, there is no such centrifugal barrier to limit contraction along the axis. This is the reason why a galaxy is disc-shaped. As for the origin of the angular momentum, it remains an important unsolved problem in cosmology.

6-11 The Very Early Universe

When the universe was younger than 1 second, the temperature was high enough to produce heavier particle-antiparticle pairs. For example, muons have a rest mass energy of about 100 MeV, which means that before $T_r = 10^{12}K$, namely, when age t was less than about $2 \times 10^{-4}s$, $\mu^+ - \mu^-$ pairs could have been produced. Protons have a rest mass energy about 10^3 MeV, so that proton-antiproton pairs could have existed before.

$$T_r \sim 10^{13}K, \qquad t \sim 2 \times 10^{-6}s \ .$$

In those first moments, there were therefore a great number of particle and antiparticle types. We know that a particle-antiparticle pair is symmetrical from particle physics. However, on the universal scale, the numbers of particles and antiparticles would not be equal for the following reason. If the numbers had been precisely equal, the annihilation after cooling would surely have been almost complete and there would be virtually no protons, neutrons and the various nuclei today. For this reason, the following question must be answered in cosmology: why was the ratio of particle and antiparticle numbers in the very early universe asymmetrical?

Another problem about the origin of particles is the ratio of photon number to particle number. Since the photon number density is $n_r \simeq aT^3/k$ and particle number density is $n \simeq \rho/m_p$, we have, from recent observations,

$$\frac{n}{n_r} \simeq 10^{-9}$$

at the present time. However, noticing that $T \simeq R^{-1}$ and $\rho \simeq R^{-3}$, it is obvious that the ratio n/n_r in the expanding universe is constant and was therefore determined by the development of the extremely early universe, which again must be explained.

We list the other problems which belong to the very early universe as follows:

1. Large scale homogeneity and isotropy. On scales larger than 100 Mpc, the universe appears isotropic. Uniformity of the microwave background in any direction in the sky suggests that the universe is also homogeneous on large scales. Homogeneity beyond the causally connected region $\sim ct$ (see section 6.10) is unexplained.

2. Flatness. The present mass density ρ of the universe is close to the critical density ρ_c which separates the open and closed universes: $0.01\rho_c < \rho < 10\rho_c$. As time t evolves, the energy density varies as $t^{-3/2}$ (matter-dominated era) or t^{-2} (radiation-dominated era), while the critical density varies only as t^{-1}. Thus, at cosmic time $t \simeq 10^{-44}$s, the energy density must be fine-tuned to the critical density to the accuracy of order 10^{-60}. A typical universe, instead, would recollapse soon after the beginning, which makes our old universe an exception among infinite possibilities.

3. Cosmological constant. From present observational data, the

cosmological constant Λ is known to be less than 10^{-44}GeV^4 times $8\pi G$ in magnitude. Lack of a symmetry which forbids this term makes the smallness of the cosmological constant a complete mystery.

The advent of gauge theories that unify strong and electroweak interactions, namely grand unified theories, provided a natural explanation of the particle-antiparticle asymmetry, and opens up new possibilities of a solution to both the homogeneity and flatness problems. We still do not have a clue to understanding the small cosmological constant. The technical aspects of these problems already go beyond the scope of this book. We can only show, in Table 19, the history of our universe as is known at present.

Table 19. The evolutionary stages of the universe

Time(s)	Temperature(K)	Energy	Physical Processes
10^{-44}	10^{32}	10^{19} GeV	Planck era
10^{-36}	10^{28}	10^{15} GeV	Origin of particle asymmetry
10^{-12}	10^{16}	10^{3} GeV	Phase transition of electroweak
10^{-4}	10^{12}	10^{2} GeV	Lepton era starts
10^{-2}	10^{11}	10 MeV	Decoupling of neutrinos
1	10^{10}	1 MeV	$e^{+}e^{-}$ annihilation
10^{2}	10^{9}	0.1 MeV	He formation
10^{12}	4×10^{3}	0.4 eV	Recombination
$\sim 4 \times 10^{17}$	2.7	3×10^{-4} eV	Present

In Table 19, the Planck era is determined by the condition that

213

quantum effects must be considered. For the very early universe, the Hubble constant is given by (6.36) and (6.40) as H(t) = 1/2t, which describes the expansion rate of the universe. On the other hand, the quantum oscillation rate is estimated by $kT_r/\hbar$. The quantum effects occur when the time scale of the expansion equals that of quantum oscillations. In this way, we find the Planck era should be earlier than t_p, which is the solution of H = $kT_r/\hbar$ i.e.

$$t_p = \frac{\pi}{3} \left(\frac{hG}{5c^5}\right)^{1/2} = 2.5 \times 10^{-44}s \ .$$

REFERENCES

General introductions and expositions on cosmology

D.W. Sciama, Modern Cosmology (1971), (Cambridge University Press).

P.I.E. Peeble, Physical Cosmology (1971), (Princeton University Press).

Concerning the apparent magnitude-redshift relation

A. Sandage, A. Tammann, Astrophysical Journal 194, 223, 559 (1974). 196, 313 (1975), 197, 265 (1975).

L.Z. Fang, T. Kiang, F.H. Cheng and F.X. Hu, Quart. J. Roy. Astron. Soc. 23, 363 (1982).

Microwave background radiation

A.A. Penzias, R.W. Wilson, Astrophysical Journal 142, 419 (1965).

D.P. Woody, J.C. Mather, N.S. Nishioka, P.L. Richards, Phys. Rev Letters 34, 1036 (1975).

The formation of elements

R.V. Wagoner, W.A. Fowler, F. Hoyle, Astrophysical Journal 148, 3 (1967).

H. Reeves, Annual Review of Astronomy and Astrophysics 12, 437 (1974).

V. Trimble, Rev. Mod. Phys. 47, 877 (1975).

abnormal neutron star   95
absolute magnitude   97, 174
absolute space   3
accretion   128
age of pulsars   106
age of the universe   187
anisotropy   198, 199
antiparticle   211
apparent magnitude   120, 170
  -redshift relation   174, 175

background black body radiation   193
baryon number   123
big-bang model   180
binary system   65, 101, 130, 146
black hole   84
  emission   125
  entropy   125
  temperature   125

caesium clock   49
Cen X-3   135, 137
Chandrasekhar limit   90
Christoffel symbol   26
closed universe   170
clusters of galaxies   183
collapse of star   114
comoving observer   168
constant curvature space   34
coordinate condition   149
cosmic ray   97
cosmic scale factor   169
cosmological constant   184, 212
cosmological principle   168, 178
Crab nebula   77
critical density   186
critical mass   111
curvature   26, 28
Cyg X-1   134, 140

deceleration parameter   172, 176
deflection of light   56

degenerate electron 90
deuterium 203
deviation equation 147
dynamics of the expanding universe 184

Eddington limit 138
Einstein equation 37
equivalence principle 14, 40, 56
ergosphere 129
event horizon 129

flatness 213
formation of galaxies 205

galaxy 1, 175
Galilean transformation 2
γ-ray pulsar 109
geocentric universe 166
geodesic 23
globular cluster 192
gravitational
  instability 206
  mass 7
  radiation damping 156, 162, 163
  radius 38, 117
  redshift 40
  wave 144
guest star 73

helium abundance 201
Her X-1 134, 135
Hertzsprung-Russell diagram 96
homogeneous universe 167
horizon 118, 123
Hubble constant 176
hyperon fluid 94

inertial frame 3
inertial mass 7
isotropic universe 167

Jeans mass 206

K correction 175
Kerr metric 122
Kerr-Newman metric 121, 122
Kronecker symbol 26

lepton number 123
local inertial frame 17
Lorentz transformation 11

Mach's principle 12, 144, 146
matter-dominated era 194
Mercury's precession 10, 14, 49, 54
metric tensor 21
microwave background radiation 196
Minkowski's metric 22, 148
missing mass 186
Mösbauer effect 42

Neptune 8, 10
neutrino 202
neutron star 84, 90, 111
Newtonian gravitation 6
"no hair" 123
nucleosynthesis 202
null geodesics 25, 179
number counts 180

oblateness of the sun 55
occultation 133
Olbers' paradox 178
open universe 170
Oppenheimer-Volkoff limit 92
origin of particle asymmetry 212

π-condensation 95
Planck era 213
polytropic index 87
post-Newtonian correction 53, 63
power spectrum 105
precession of the axis of rotation 65
precession of perihelion 49, 65
proper time 19
pulsar 96

QSO, see quasar
quadrupole 153
quasar 60, 177

radar echoes 61
radiation-dominated era 194
recombination time 201
Reissner-Nordstrom metric 122
relativity 11, 15
Riemannian tensor 33, 148
Robertson-Walker metric 169
Roche lobe 131, 133

Schwarzschild metric 44
separation 21
singularity 116, 187

standard candle   176
stress-energy-momentum tensor   37
supernova   74, 161
synchrotron radiation   80

temperature of background radiation   194
tidal force   17, 147
time keeping   158
twin paradox   23
two-body problem   65

uranium   193

white dwarf   83, 88, 96

X-ray binary   134
X-ray pulsar   135